STUDENT SOLUTIONS MANUAL TO ACCOMPANY

INTRODUCTION TO COORDINATION, SOLID STATE, AND DESCRIPTIVE INORGANIC CHEMISTRY

STUDENT SOLUTIONS MANUAL TO ACCOMPANY

INTRODUCTION TO COORDINATION, SOLID STATE, AND DESCRIPTIVE INORGANIC CHEMISTRY

Glen E. Rodgers

Allegheny College

McGraw-Hill, Inc.

New York St. Louis San Francisco Auckland Bogotá Caracas Lisbon
London Madrid Mexico City Milan Montreal New Delhi
San Juan Singapore Sydney Tokyo Toronto

STUDENT SOLUTIONS MANUAL TO ACCOMPANY RODGERS:
INTRODUCTION TO COORDINATION, SOLID STATE, AND DESCRIPTIVE INORGANIC CHEMISTRY

This book is printed on recycled paper containing 10% post consumer waste.

2 3 4 5 6 7 8 9 0 MAL MAL 9 0 9 8 7 6 5 4

ISBN 0-07-053389-X

The editor was Jennifer Speer;
the production supervisor was Diane Ficarra.
Malloy Lithographing, Inc., was printer and binder.

Table of Contents

	Preface	vii
Chapter 1	The Evolving Realm of Inorganic Chemistry	1
Chapter 2	An Introduction to Coordination Chemistry	2
Chapter 3	Structures of Coordination Compounds	7
Chapter 4	Bonding Theories for Coordination Compounds	15
Chapter 5	Rates and Mechanisms of Reactions of Coordination Compounds	25
Chapter 6	Applications of Coordination Compounds	33
Chapter 7	Solid State Structures	38
Chapter 8	Solid State Energetics	47
Chapter 9	Building a Network of Ideas to Make Sense of the Periodic Table	56
Chapter 10	Hydrogen and Hydrides	62
Chapter 11	Oxygen, Aqueous Solutions, and the Acid Base Character of Oxides and Hydroxides	68
Chapter 12	Group 1A: The Alkali Metals	76
Chapter 13	Group 2A: The Alkaline Earth Metals	81
Chapter 14	The Group 3A Elements	87
Chapter 15	The Group 4A Elements	94
Chapter 16	Group 5A: The Pnicogens	100
Chapter 17	Sulfur, Selenium, Tellurium, and Polonium	107
Chapter 18	Group 7A: The Halogens	115
Chapter 19	Group 8A: The Noble Gases	122

Preface

In each chapter of this *Student's Solutions Manual* you will find the following three items: (1) a listing of the sections and subsections, (2) the objectives of each chapter, and (3) the answers to the odd-numbered end-of-the-chapter problems.

Solutions manuals always have some mistakes. While every effort has been made to minimize their number, there inevitably will be some. I would appreciate it if you would let me know when you find mistakes by writing to me at the Department of Chemistry, Allegheny College, Meadville, PA. Thanks in advance for your help.

The author is indebted to the students in his sophomore inorganic courses over the last 10 years for their many insightful questions and discussions regarding many of these problems and their solutions. This manual was prepared on a NeXT computer using WordPerfect for NeXT and !Diagram. The author is grateful to Brenda Metheny for her help in preparing the manuscript and to the staff of the Educational Computing Services of Allegheny College for quick and efficient trouble shooting when occasional problems arose.

Glen E. Rodgers

INTRODUCTION TO COORDINATION, SOLID STATE, AND DESCRIPTIVE INORGANIC CHEMISTRY

Chapter 1
The Evolving Realm of Inorganic Chemistry

Chapter Objectives

You should be able to

- □ explain what the realm of inorganic chemistry encompasses
- □ appreciate when chemistry first became recognized as a separate academic discipline
- □ appreciate when and how inorganic chemistry became established as a subdiscipline of chemistry
- □ appreciate the role and some of the accomplishments of inorganic chemists in the latter half of the nineteenth century
- □ explain the role of the quantum revolution in changing the nature of inorganic chemistry
- □ appreciate some of the major accomplishments of inorganic chemists in the first half of the twentieth century
- □ appreciate some of the major accomplishments of inorganic chemists in the second half of the twentieth century

Chapter 2
An Introduction to Coordination Chemistry

The sections and subsections in this chapter are listed below.

2.1 The Historical Perspective
2.2 The History of Coordination Compounds
  Early Compounds
  The Blomstrand-Jørgensen Chain Theory
  The Werner Coordination Theory
2.3 The Modern View of Coordination Compounds
2.4 An Introduction to the Nomenclature of Coordination Compounds

Chapter Objectives

You should be able to

- define some important terms used in coordination chemistry
- give a few examples of coordination compounds encountered in earlier courses
- put coordination chemistry into the historical context of the conceptual development of atomic structure, the periodic table, and chemical bonding
- relate how the formulas and properties of early coordination compounds were but incompletely rationalized by the Blomstrand-Jørgensen chain theory
- explain how Werner's coordination theory, with its concept of primary and secondary valences, more completely rationalized the properties of early coordination compounds
- draw structural formulas for coordination compounds using both the Blomstrand-Jørgensen chain theory and the Werner coordination theory
- explain how Werner established that the secondary valence of cobalt(III) is directed to the corners of an octahedron
- work with a variety of coordination compounds involving monodentate, multidentate, bridging, and ambidentate ligands
- name coordination compounds involving a variety of metals, ligands, and counterions

2.1. Dalton's atomic theory had three essential components. First, he proposed that all elements were composed of tiny, indivisible particles called atoms. All the atoms of a given element were the same in every way while those of different elements were different in properties such as mass and size. Second, compounds were composed of atoms of more than one element in small, whole number ratios, 1:1, 2:1, etc. Third, a chemical reaction involved the shuffling of atoms from one compound to another. The idea of hooks embedded into atoms could have been a way to account for the number of other atoms with which a given atom could combine. That is, an atom of oxygen, which often combines with two other atoms of a second element in such compounds as water, H_2O, might have two embedded hooks while hydrogen atoms might have only one.

2.3. In carbon monoxide, the valence of carbon is one while in carbon dioxide it is two. Therefore, these two compounds are not consistent with the concept of fixed valence. In practice, it is found that a given element may have several characteristic valences.

2.5. The quantum mechanical atom pictures an electron in an atom to occupy only certain, allowed energy levels. When this atom is excited, that is, when energy is put into it, the electron can be moved from one allowed level to a higher one. Once the atom moves away from the energy source, the electron will return to its lower "ground state" energy level. The energy lost, E, according to Planck's equation, $E = h\nu$, would correspond to a specific frequency or wavelength of light emitted. As various electrons are excited and then return to their lower levels, a set of energies (and therefore a set of wavelengths), characteristic of the particular atom would be emitted. This characteristic set of wavelengths is known as an emission spectrum.

2.7. Mendeleev used valence to determine where he should leave blanks in his periodic table. He always tried to put elements with similar valences in the same column or group. However, sometimes the next element in his sequence by atomic weight did not have the same valence as the others in the next column or columns. In these cases, he left a blank in the table and put the next known element in the following appropriate group. For example, after zinc, the next known element was arsenic, but it had a valence of five not three as dictated by its placement under aluminum or four if it were under silicon. Accordingly, Mendeleev in this case skipped two groups saying that there were two as yet undiscovered elements in groups 3 and 4 under aluminum and silicon, respectively. These elements he called eka-aluminum and eka-silicon, respectively. They were discovered shortly thereafter and named gallium and germanium.

2.9. The two isomers of $CoCl_3 \cdot 4NH_3$ might be

```
  Cl                                   NH3-NH3-NH3-NH3-Cl
 /                                    /
Co-NH3-NH3-NH3-NH3-Cl               Co- Cl
 \                                    \
  Cl                                   Cl
```

2.11. (a) Blomstrand-Jørgensen:

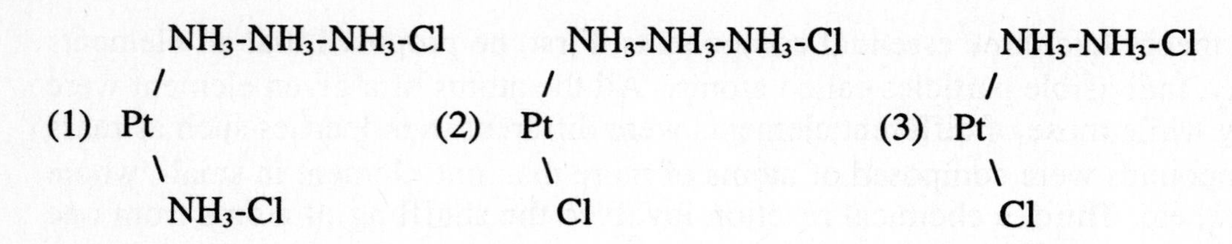

Note that the conductivity and number of chloride ions that could be precipitated by $AgNO_3$ are correctly predicted for compounds (1) and (2) by the Blomstrand-Jørgensen theory. However, it does not account for the properties of compound (3).

(b) Werner Formulas:

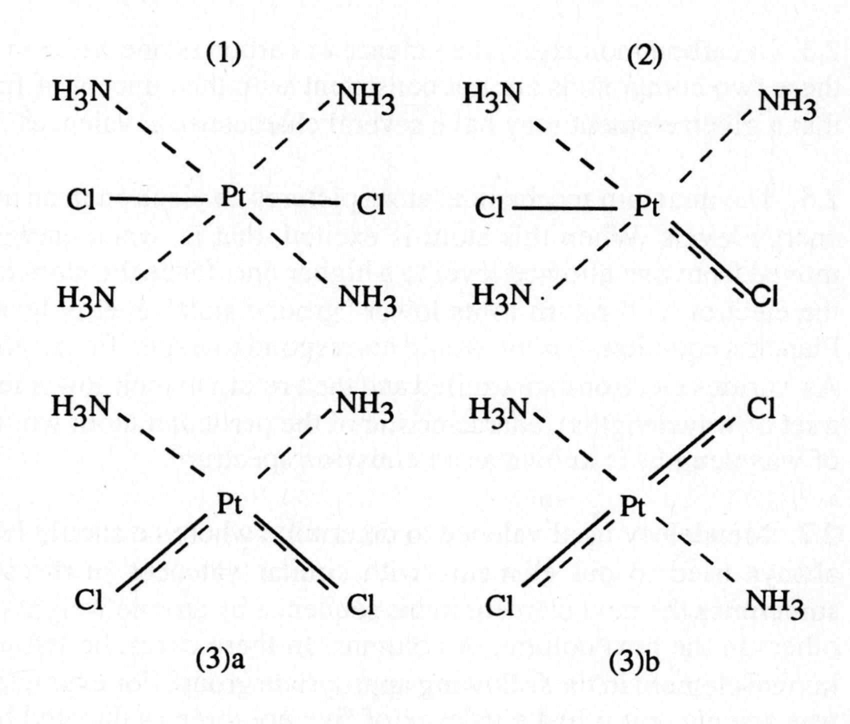

(c) Modern Structural Formulas:

(1) $[Pt(NH_3)_4]Cl_2$
(2) $[Pt(NH_3)_3Cl]Cl$
(3) $[Pt(NH_3)_2Cl_2]$

2.13. MA_2B_2

Square Planar: two possible isomers.

Tetrahedral: only one possible isomer because all the possible positions are adjacent to each other.

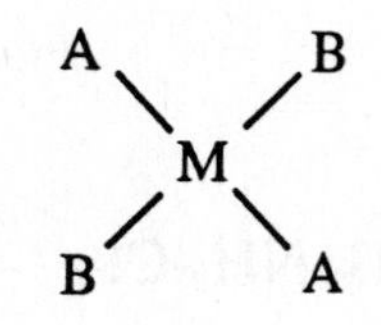

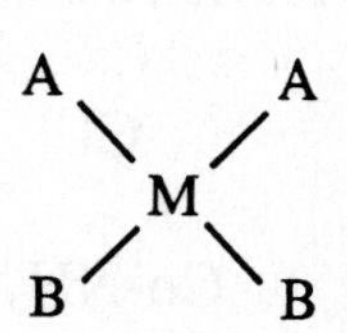

If $Pt(NH_3)_2Cl_2$ has two possible isomers it is square planar. With only one possible isomer, $[CoBr_2I_2]^2$ must be tetrahedral.

2.15. There would be three possible isomers for a hexagonal planar secondary valence.

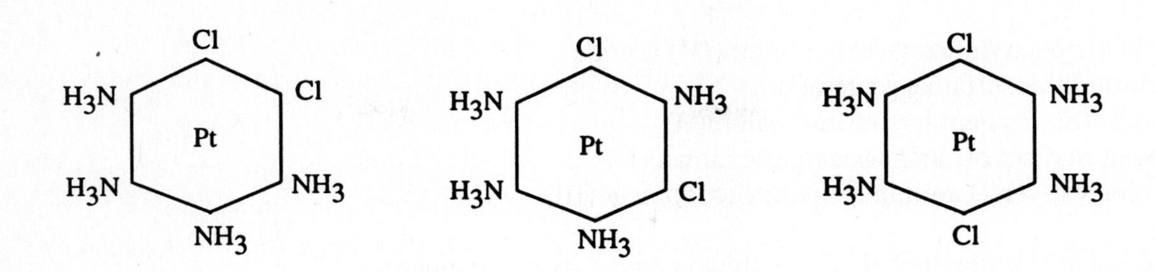

2.17. Tricarbonyltrichlorochromium(III)

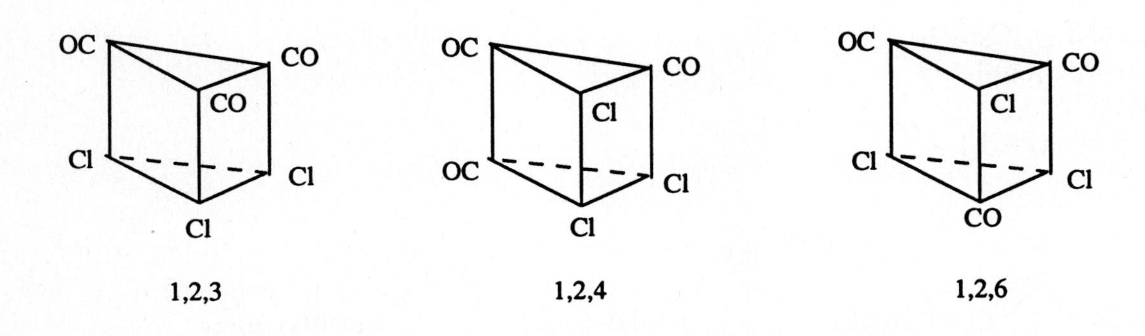

2.19. Ambi- is a combining prefix meaning "both." Ambidextrous, therefore, means able to use both hands while ambidentate, where dentate is used to indicate a ligand binding to a metal atom, means that the ligand has two binding sites. A difference in these two terms is that ambidextrous is usually taken to indicate an ability to use both hands with *equal ease.* No such equality is implied in the term ambidentate.

2.21. Only ammonia and nitrite are possible ligands as they are the only ones to have lone pairs that can be donated to a metal ion.

Formula	No. of ionic nitrites	No. of geometric isomers
$[Co(NH_3)_6](NO_2)_3$	3	1
$[Co(NH_3)_5(NO_2)](NO_2)_2$	2	1
$[Co(NH_3)_4(NO_2)_2]NO_2$	1	2
$[Co(NH_3)_3(NO_2)_3]$	0	2
$K[Co(NH_3)_2(NO_2)_4]$	0	2
$K_2[Co(NH_3)(NO_2)_5]$	0	1
$K_3[Co(NO_2)_6]$	0	1

2.23. (a) tetraamminedichloroplatinum(IV) sulfate
(b) potassium hexacyanodifluoromolybdenate(V)
(c) potassium (ethylenediaminetetraacetato)cobaltate(III)

(d) triamminetrinitrocobalt(III)
(e) tris(ethylenediamine)iron(III) hexachloroiridate(III)

2.25. (a) tetrakis(triphenylphosphine)platinum(IV) acetate
(b) calcium dithiosulfatoargentate(I)
(c) dibromotris(triphenylarsine)ruthenium(II)
(d) potassium diaquo(nitrilotriacetato)cadmate(II)
(e) diamminesilver(I) amminepentanitritocobaltate(III)

2.27. (a) bis(ethylenediamine)cobalt(III)-μ-thiocyanato-μ-isothiocyanato-bis(acetylacetonato)chromium(III) nitrate
(b) triaquocopper(I)-μ-diacetatotriaquocopper(I)
(c) diaquotris(thiocyanato)chromium(III)-μ-hydroxopentaamminecobalt(III) sulfate

2.29. (a) $[Ag(CH_3NH_2)_2]C_2H_3O_2$
(b) $Ba_3[CoBr_2(C_2O_4)_2]_2$
(c) $NiCO[P(C_6H_5)_3]_3$
(d) $[Pt(C_5H_5N)_4][PtCl_4]$

2.31.

(a)

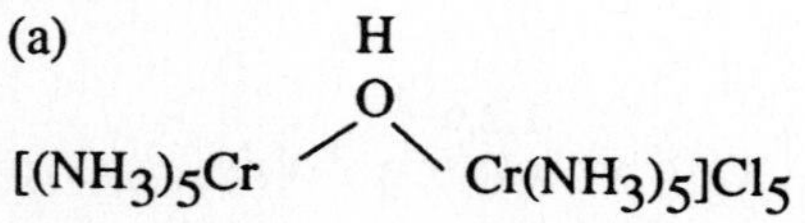

H
O
$[(NH_3)_5Cr$ / \ $Cr(NH_3)_5]Cl_5$

(b)

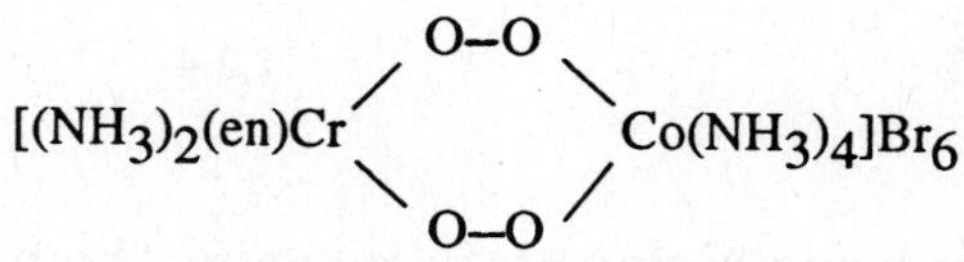

O–O
$[(NH_3)_2(en)Cr$ / \ $Co(NH_3)_4]Br_6$
O–O

Chapter 3
Structures of Coordination Compounds

The sections and subsections of this chapter include

- 3.1 Stereoisomers
- 3.2 Octahedral Coordination Spheres
  - Compounds with Monodentate Ligands
  - Compounds with Chelating Ligands
- 3.3 Square Planar Coordination Spheres
- 3.4 Tetrahedral Coordination Spheres
- 3.5 Other Coordination Spheres
- 3.6 Structural Isomers

Chapter Objectives

You should be able to

- ▫ distinquish among the various types of stereoisomers and structural isomers
- ▫ define, describe, test for, and be familiar with the nomenclature of chirality
- ▫ determine the number and types of and name stereoisomers commonly encountered in octahedral coordination compounds
- ▫ describe the general approach to resolving isomers of ionic coordination compounds
- ▫ demonstrate how Werner forever laid to rest the idea that chirality was a property associated only with carbon
- ▫ determine the number and types of and name stereoisomers commonly encountered in square planar and tetrahedral coordination compounds
- ▫ appreciate and give examples of coordination compounds with coordination numbers other than four or six
- ▫ define and give an example of fluxional five-coordinate compounds
- ▫ recognize and predict the incidence of coordination and ionization structural isomers
- ▫ describe, name, and predict the incidence of linkage isomers that occur with the common ambidentate ligands

3.1. Ethanol and dimethyl ether are isomers because they have the same number and types of atoms but different properties. Furthermore, they are structural isomers because they have different numbers and types of chemical bonds. For example, ethanol has one C-O bond while dimethyl ether has two.

3.3. The two chain theory formulations would be considered to be stereoisomers because they have the same numbers and types of chemical bonds (three each of Co-N, NH_3-NH_3, and NH_3-Cl), but differ in the spatial arrangements of those bonds.

3.5. A good pair of scissors does not generally have an internal mirror plane and therefore is chiral. Left-handed people often find themselves using scissors made for the majority of the population which is right-handed.

3.7. Each of these sets of isomers (the 1,2; 1,3; and 1,4 for the MA_4B_2 and the 1,2,3; 1,2,4; and 1,3,5 for the MA_3B_3) are geometric isomers because their different spatial arrangements result in different geometries. None of these geometric isomers are chiral because each contains at least one internal mirror plane, namely that of the hexagon itself.

3.9. The two mirror images, shown at right, can be superimposed upon each other. Therefore, this molecule is not chiral. The same conclusion is arrived at by considering that the molecule has at least one internal mirror plane.

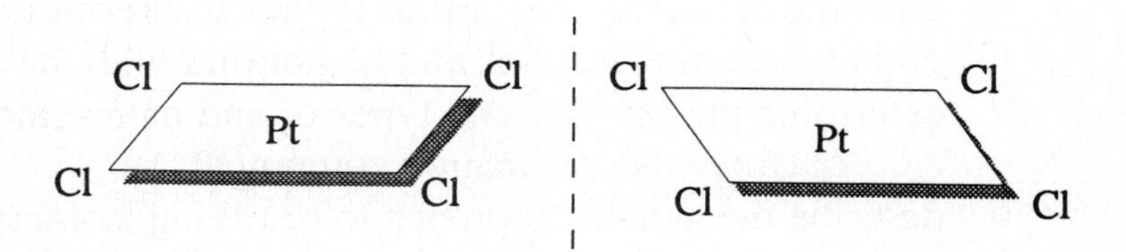

3.11. These two mirror images are nonsuperimposable. No matter how one is rotated in space, it cannot be made equivalent to the other. Therefore, this molecule is chiral. The same conclusion is arrived at by considering the fact that this molecule does not contain any internal mirror planes.

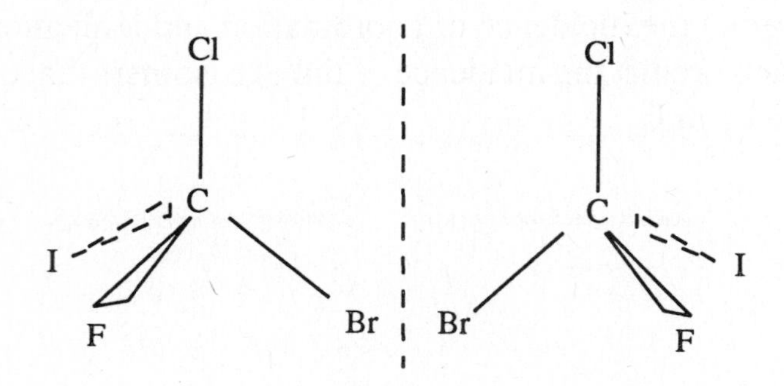

3.13. (a) If the mirror image of a molecule cannot be rotated in space so as to be equivalent to the original molecule, it is said to have a nonsuperimposable mirror image. Such a molecule is chiral. (b) If a molecule does not possess an internal mirror plane (a plane that passes through the molecule such that every atom in the molecule can be reflected through the plane into another equivalent atom) it is chiral. [There are a few exceptions to this rule but they are beyond the scope of this text.]

3.15. The structures of these two molecules are shown below. Only AlClBrI has an internal mirror plane; the one that the entire molecule sits in. Therefore, this molecule is not chiral whereas the corresponding phosphorus compound is.

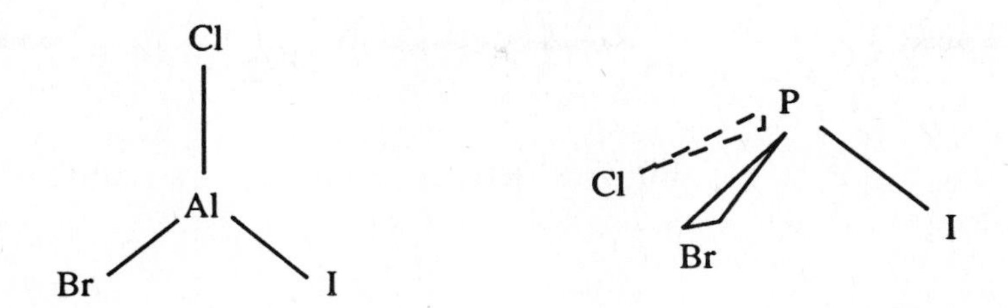

3.17. The structural formulas are as follows: [For (b) and (c), the structure of only the cations are shown; the overall +1 positive charge of these two cations is also omitted for clarity.]

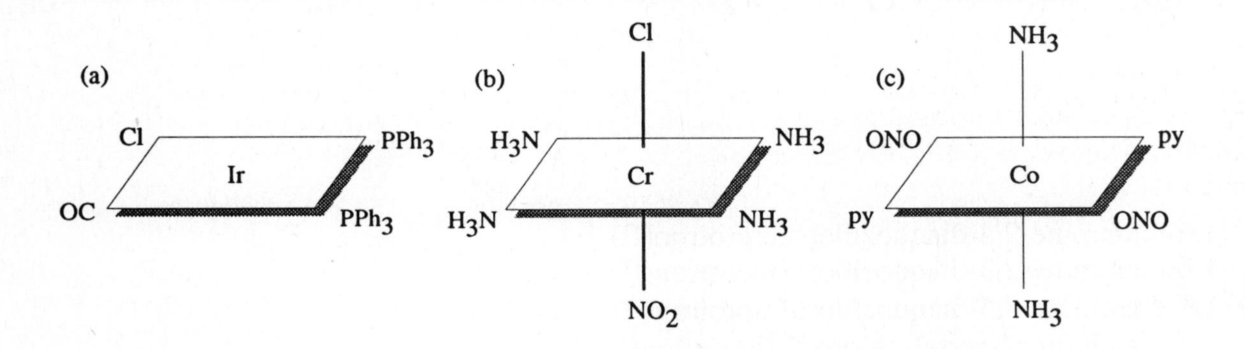

3.19.

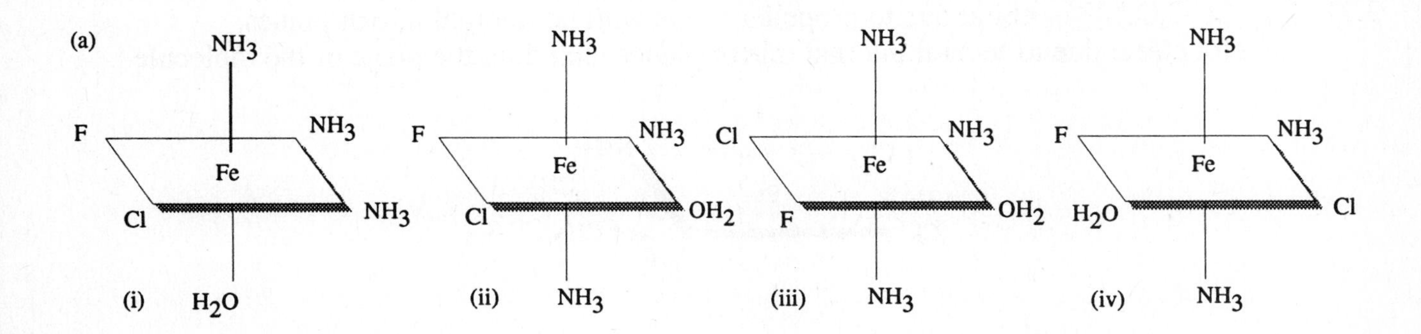

(i) 1,2,3-triammineaquochlorofluoroiron(II)
(ii) 1,2,6-triammine-3-aquo-4-chlorofluoroiron(II)
(iii) 1,2,6-triammine-3-aquochloro-4-fluoroiron(II)
(iv) 1,2,6-triammine-4-aquo-3-chlorofluoroiron(II)

3.19. (cont.)

(b)

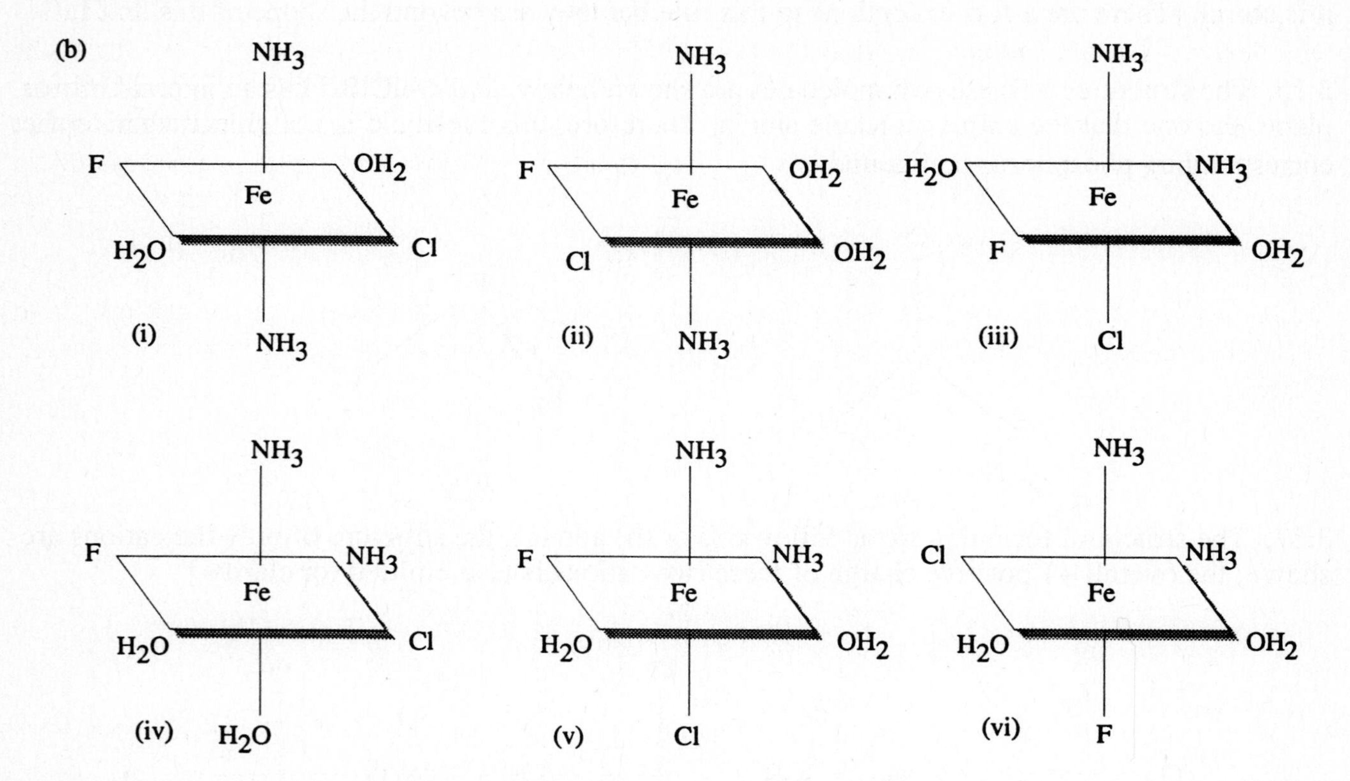

(i) 1,6-diammine-2,4-diaquochlorofluoroiron(II)
(ii) 1,6-diammine-2,3-diaquochlorofluoroiron(II)
(iii) 1,2-diammine-3,5-diaquochlorofluoroiron(II)
(iv) 1,2-diamminediaquo-3-chloro-5-fluoroiron(II)
(v) 1,2-diammine-3,4-diaquochloro-5-fluoroiron(II)
(vi) 1,2-diammine-3,4-diaquo-5-chlorofluoroiron(II)

3.21. (a) $[Cr(C_2O_4)_3]^{3-}$ -- chiral due to propeller shape with no internal mirror planes.
(b) Not chiral due to several internal mirror planes including the plane of the molecule.

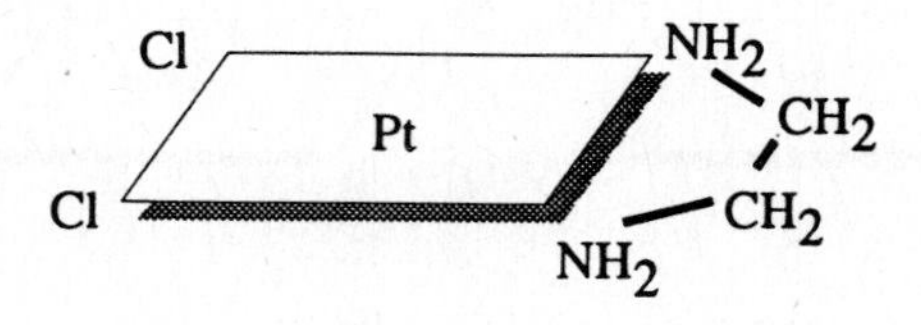

3.23. Prior to Werner's synthesis and resolution of coordination compounds containing chelating agents, the evidence for his coordination theory had been "negative." That is to say, his theory often predicted only two isomers for a given MA_4B_2 or MA_3B_3 complex (where A and B are monodentate ligands) if the geometry of the coordination sphere was octahedral. Planar hexagonal and trigonal prismatic geometries, on the other hand, yielded three isomers in each case. His and other research groups indeed could synthesize only two isomers of these compounds. But what if the correct geometry was hexagonal or trigonal prismatic and the third isomer was just particularly difficult to synthesize? Werner had only "negative" evidence for his octahedral coordination theory. It was the *absence* of a given isomer that Werner claimed supported his ideas. However, octahedral coordination spheres for a complex of general formula $M(AA)_2B_2$ (where AA = a chelating bidentate ligand) should have one chiral and one nonchiral form. Werner and his group were able to confirm the existence of this number of isomers and therefore provide "positive" evidence for his theory and the octahedral configuration of his secondary valence.

3.25. For the complex cation of this compound (shown without its charge), there would be four possible geometric isomers as shown below. Only the last, (d), lacks an internal mirror plane and is chiral.

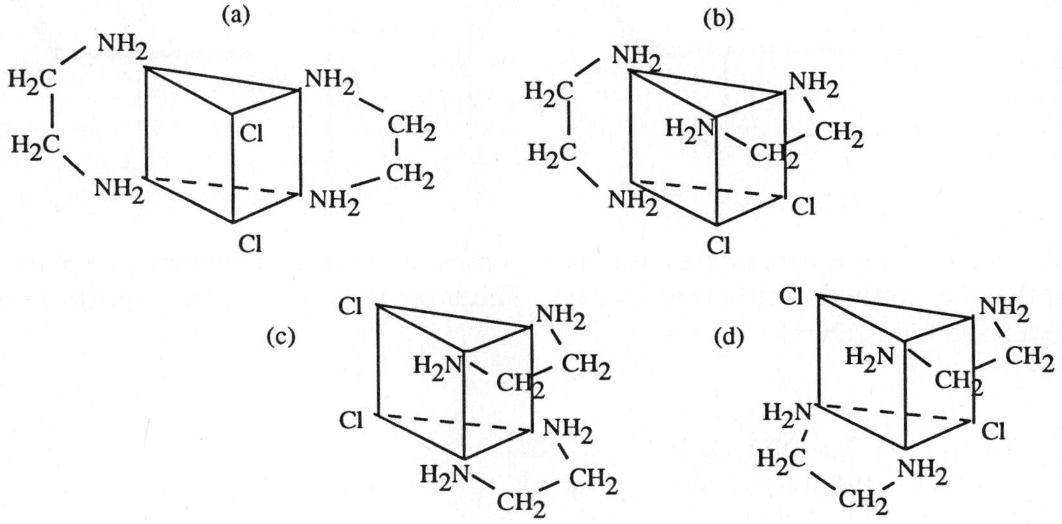

3.27. Since both the ethylenediamine and oxalate ligands can only span the *cis* positions of an octahedron, there is only one possible geometric isomer for this complex. This cation (shown below without its +1 charge) lacks a plane of symmetry and is chiral.

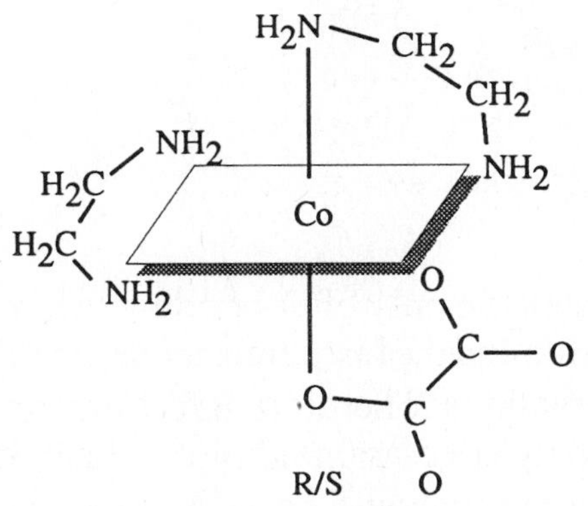

3.29. (a) potassium chloro(nitrilotriacetato)thiocyanatocobaltate(III)
(b) Since the four binding cites of the NTA must be adjacent to each other, the two monodentate ligands (SCN^- and Cl^-) also must always be *cis* to each other. However, either the chloride or the thiocyanate could be *trans* to the nitrogen of the NTA as shown below. In either case, the complex anion contains an internal mirror plane and is therefore not chiral.

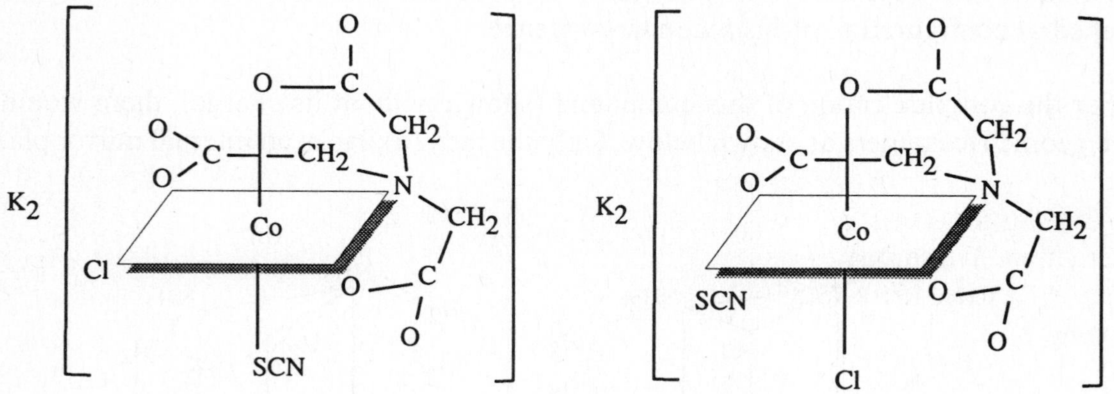

K_2 [Co(NTA)(SCN) Cl] — Cl in plane, SCN axial

K_2 [Co(NTA)(SCN) Cl] — SCN in plane, Cl axial

3.31. Both of the square planar structures possess a plane of symmetry (the plane of the molecule), but the one tetrahedral structure does not. Therefore the square planar versions are nonchiral while the tetrahedral is chiral.

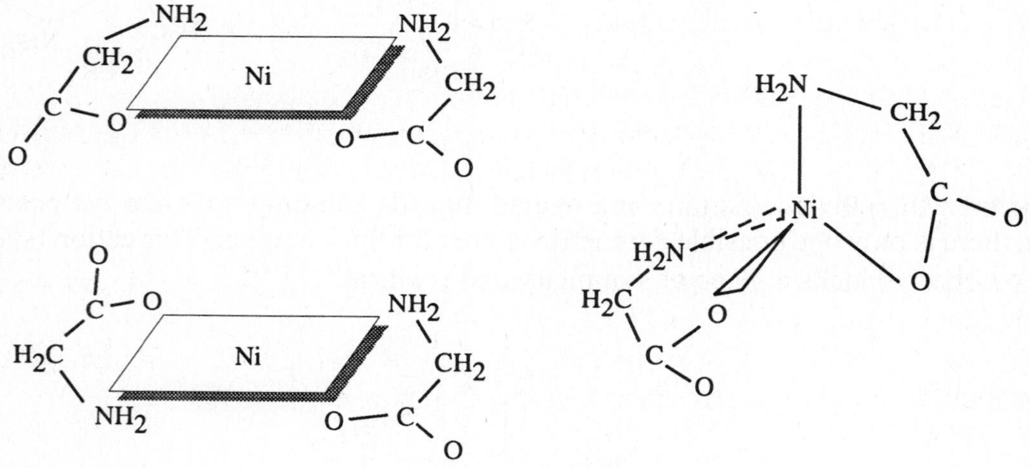

3.33. Since the nitrate ligand is capable only of O-bonding to a metal center, both the bis(ethylene-diamine)dinitratocobalt(III) chloride and the bis(ethylenediamine)dinitritocobalt(III) chloride possess four Co^{3+}-N and two Co^{3+}-O interactions. Therefore it follows that both of these compounds should be the same color while the bis(ethylenediamine)dinitrocobalt(III) chloride, in which the complex cation contains six Co^{3+}-N interactions, should be a different color.

3.35. The Lewis and VSEPR structures of thiocyanate are shown in (i) and (ii) below. The sulfur has three lone pairs available to bind with a metal center while the nitrogen has only one. Assuming the four electron pairs around the sulfur (three lone pairs and one bonding pair) are tetrahedrally dispersed, S-bonded SCN^- complexes should possess a nonlinear M-S-C interaction. Given the linear arrangement of the triple bond and the one lone pair about the nitrogen atom, on the other hand (or is it on the other tooth?) N-bonded thiocyanate complexes should possess a linear M-N-C interaction. These two cases are shown in structures (iii) and (iv), respectively.

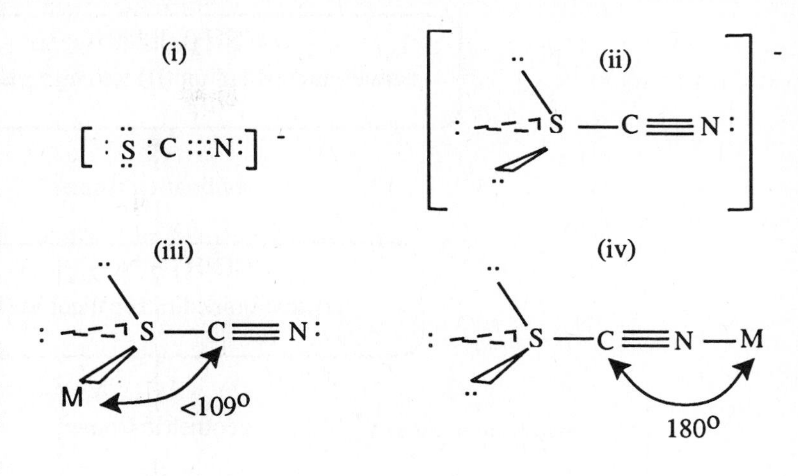

3.37. S-bonded thiocyanate will take up more space around a metal atom or ion because the -SCN is able to rotate about the M-S bond and sweep out a fairly large cone-shaped volume. N-bonded thiocyanate, on the other hand (tooth?), rotates about the M-N bond and sweeps out a much smaller, needle-like cylindrical volume. In the $[Co(NH_3)_5NCS]^{2+}$ cation, the five ammonia ligands sweep out cone-like volumes and take up a fair amount of volume. They do not leave much volume available for the thiocyanate so it is more stable in the isothiocyanate "needle" form that does not require as much volume. In the $[Co(CN)_5SCN]^{3-}$ anion, the linear CN^- ligands take very little volume and so the S-bonded SCN^-, that takes up more space, is favored.

3.39. $[Cu(NH_3)_4][PtBr_4]$, tetraamminecopper(II) tetrabromoplatinate(II)
$[Cu(NH_3)_3Br][Pt(NH_3)Br_3]$, triamminebromocopper(II) amminetribromoplatinate(II)
$[Pt(NH_3)_3Br][Cu(NH_3)Br_3]$, triamminebromoplatinum(II) amminetribromocuprate(II)
$[Pt(NH_3)_4][CuBr_4]$, tetraammineplatinum(II) tetrabromocuprate(II)

3.41. (a) $[Pt(NH_3)_4Cl_2]Br_2$: An ionization isomer would be $[Pt(NH_3)_4Br_2]Cl_2$, dibromotetraammine-platinum(IV) chloride.
(b) $[Cu(NH_3)_4][PtCl_4]$: A coordination isomer would be $[Cu(NH_3)_3Cl][Pt(NH_3)Cl_3]$, triamminechlorocopper(II) amminetrichloroplatinate(II).

3.43. At least one of each type of isomer except optical is possible as shown below.

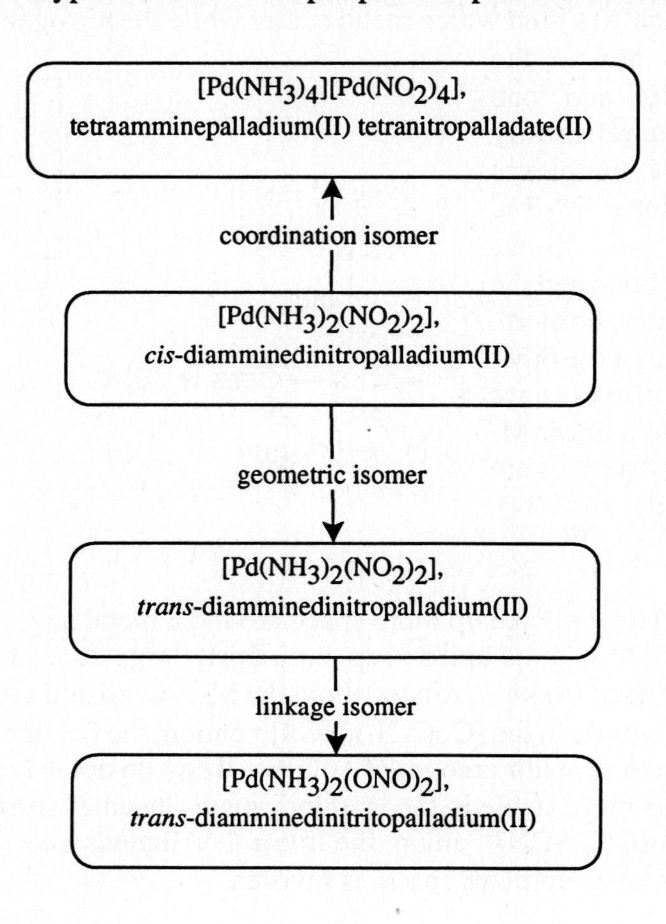

Chapter 4
Bonding Theories for Coordination Compounds

The sections and subsections in this chapter are listed below.

4.1 Early Bonding Theories
Lewis Acid-Base Theory
The Crystal Field, Valence Bond, and Molecular Orbital Theories
4.2 The Crystal Field Theory
The Shapes of the 3d Orbitals
Octahedral Fields
Tetragonally-Distorted Octahedral and Square Planar Fields
Tetrahedral Fields
4.3 Consequences and Applications of Crystal Field Splitting
Crystal Field Splitting Energies versus Pairing Energies
Crystal Field Stabilization Energies
Factors Affecting the Magnitude of the Crystal Field Splitting Energies
Magnetic Properties
Absorption Spectroscopy and the Colors of Coordination Compounds

Chapter Objectives

You should be able to

- □ explain how the Lewis acid-base theory applies to coordination compounds
- □ define and use the effective atomic number rule for coordination compounds
- □ briefly outline how the valence bond, molecular orbital, and crystal field theories generally are applied to coordination compounds
- □ qualitatively sketch out the shapes of the five independent 3d orbitals
- □ explain how the two dependent $3d_{z^2-y^2}$ and $3d_{z^2-x^2}$ orbitals are related to the $3d_{z^2}$ orbital
- □ explain in detail how the five independent 3d orbitals split in an octahedral field
- □ explain in detail the changes in the relative energies of the 3d orbitals when the z-axis ligands of an octahedral field are gradually withdrawn to produce a tetragonally elongated and, ultimately, a square planar crystal field
- □ explain in detail how the five independent 3d orbitals split in a tetrahedral field
- □ describe the strong- and weak-field, low- and high-spin cases as applied to coordination compounds
- □ calculate the crystal field stabilization energy (CFSE) in terms of the crystal field splitting (Δ) and pairing (P) energies for any given d^n case in an octahedral, tetrahedral, or square planar crystal field
- □ explain how factors such as the type of field, the size and charge of the metal ion, and the size and charge of the ligand should affect the magnitude of the crystal field splitting energy assuming a completely ionic M-L interaction
- □ rationalize the degree to which the spectrochemical series can and cannot be explained by the crystal field theory
- □ explain how admitting a certain degree of covalent character to the M-L interaction can be used to better rationalize the spectrochemical series
- □ sketch out how molar susceptibility is measured and how its value can be related to the magnetic moment, μ, and to the number of unpaired electrons in a coordination compound
- □ rationalize why so many coordination compounds are highly colored and how these colors can, in some cases, be simply related to the size of crystal field splitting energy

4.1. Hydrogen ions:

(1) Arrhenius acid — treats acids as substances that produce H^+ in solution
(2) Brønsted-Lowry acid — capable of transferring a proton (H^+)
(3) Lewis acid — H^+ is an electron pair acceptor

Hydroxide ions:

(1) Arrhenius base — treats bases as substances that produce OH^- in solution
(2) Brønsted-Lowry base — capable of accepting a proton (H^+)
(3) Lewis base — OH^- is an electron pair donor

4.3. The oxygen molecule has a number of lone pairs of electrons associated with it. One of these lone pairs can be donated to the central iron ion (Fe^{2+}) of the hemoglobin. The O_2, then, has served as an electron pair donor and is a Lewis base. The Fe^{2+} is an electron pair acceptor and is a Lewis acid.

4.5.	Metal electrons	Ligand electrons	Total electrons	Result
(a)	Ir^{3+} 74	6 Cl^- 12	86	follows EAN
(b)	Fe^0 26	5 CO 10	36	follows EAN
(c)	Cr^0 24	6 CO 12	36	follows EAN
(d)	Co^{3+} 24	2 NH_3 4	36	follows EAN
		4 NO_2^- 8		
(e)	Ru^{2+} 42	2 Cl^- 4	52	does not follow EAN
		3 PPh_3 6		

4.7. Coulomb's law says that the potential energy (PE) is equal to Q_1Q_2/r. If r is infinite, then the PE is zero. But we also recognize that the proton and the electron are of opposite charges and, therefore, the net energy will be negative for finite values of r. When the proton and electron are merged to form a hydrogen atom, the distance r between the two particles decreases and the value of Q_1Q_2/r becomes more and more negative. As the energy of the proton-electron system becomes more negative, it follows that the system must be releasing energy to the environment.

4.9. Note that both of these orbitals are located in xy plane. However, the lobes of the $3d_{xy}$ point in between the axes whereas those of the $3d_{x^2-y^2}$ point right along them. (The dashed lines are nodes.)

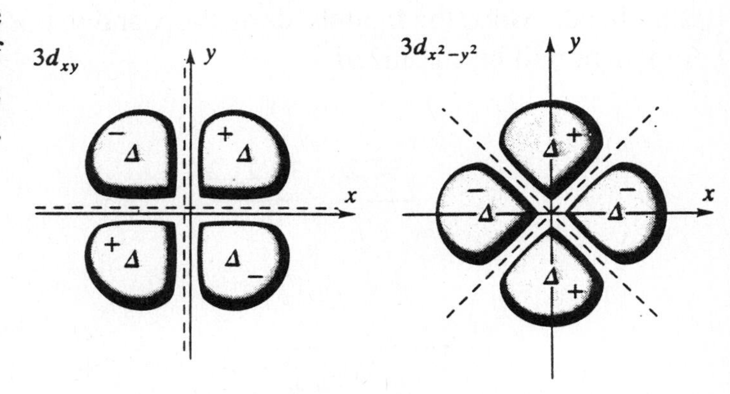

4.11.

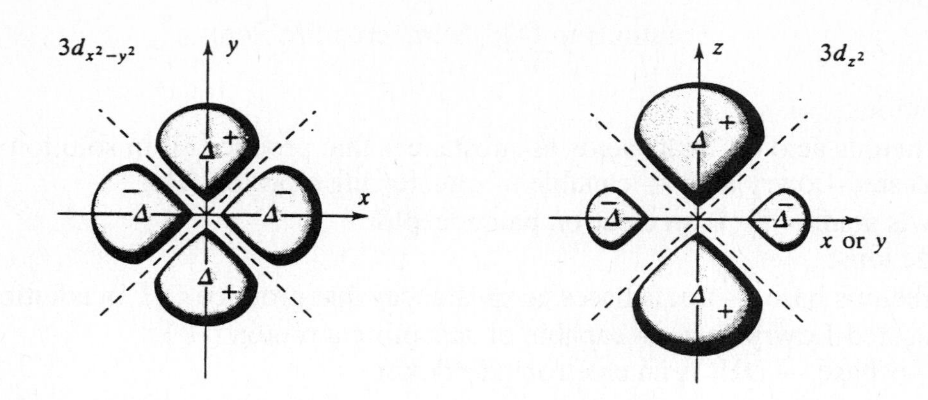

4.13. All six of the dependent d orbitals have the same shape but different orientations in space. The $3d_{z^2}$ orbital is a linear combination of the dependent $3d_{z^2-y^2}$ and $3d_{z^2-x^2}$ orbitals. Both of these latter orbitals look exactly like the other four dependent d orbitals but the $3d_{z^2-y^2}$ points along the z and y axes while the $3d_{z^2-x^2}$ points along the z and x axes. When these two orbitals are combined, the resulting $3d_{z^2}$ orbital has twice as much probability of finding an electron along the z axis as it does along either the x or the y axis.

4.15. Since each of the 3p orbitals points along a given x, y, or z cartesian axis, each would be equally affected by the two ligands of the octahedral field it would encounter. Therefore, since each 3p orbital would be affected in the same way, these orbitals would remain degenerate in such a field.

4.17. During stages I to III of the construction of an octahedral field about a metal ion, the system is becoming more and more ordered. (The twelve electrons, scattered originally an infinite distance from the metal, are brought in and organized to form the field.) A more ordered system is less random and therefore of lower entropy. It follows that entropy decreases during the construction of the field about the metal ion. If ΔS is negative for this process, ΔH must be negative in order to make ΔG negative. Recall that a reaction or process is spontaneous only if ΔG, which is equal to $\Delta H - T\Delta S$, is negative.

4.19. In a tetragonal compression along the z axis, any orbital with a z component will be destabilized. Since the ligands along the x and y axis move out slightly, any orbital with an x or y component will be stabilized.

$3d_{z^2}$ $3d_{x^2-y^2}$ ········ $3d_{z^2}$ ——

$3d_{x^2-y^2}$ ——

$3d_{xz}$ —— —— $3d_{yz}$

$3d_{xy}$ $3d_{yz}$ $3d_{xz}$ ········ $3d_{xy}$ ——

4.21.

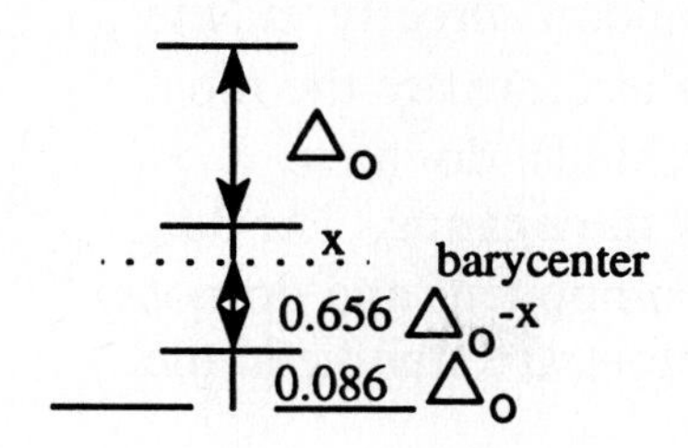

$$E_{stabilization} = E_{destabilization}$$

$$1(0.656\Delta_o - x) + 2[(0.656\Delta_o - x) + (0.086\Delta_o)] = x + (\Delta_o + x)$$

$$2.05\Delta_o - 3x = 2x + \Delta_o$$

$$1.05\Delta_o = 5x$$

$$x = 0.21\Delta_o$$

4.23.

e_g $3d_{z^2}$ $3d_{x^2-y^2}$

$3d_{x^2-y^2}$

Extra energy of stabilization due to tetragonal elongation

$3d_{z^2}$

$3d_{xy}$

t_{2g} $3d_{xy}$ $3d_{yz}$ $3d_{xz}$

$3d_{yz}$ $3d_{xz}$

Given that the barycenters of both the e_g and t_{2g} sets stay constant, there is no net gain or loss of energy when the t_{2g} set is split by the tetragonal elongation. In the e_g set, however, two electrons are stabilized by the indicated amount while only one electron is destabilized by that amount. The result is an additional energy of stabilization as indicated in the above figure.

4.25.

$3d_{xy}$ $3d_{yz}$ $3d_{xz}$

barycenter Δ_c $2/5\,\Delta_c$

$3/5\,\Delta_c$

$3d_{z^2}$ $3d_{x^2-y^2}$

4.27. The $3d_{z^2-x^2}$ and $3d_{z^2-y^2}$ orbitals point directly at the ligands located along the z axis and therefore are the most destabilized. The $3d_{yz}$ and $3d_{xz}$ orbitals do have a z-component but do not point directly at the ligands. The $3d_{xy}$ and $3d_{x^2-y^2}$ orbitals do not have a z-component and do not point directly at the ligands and therefore are stabilized the most.

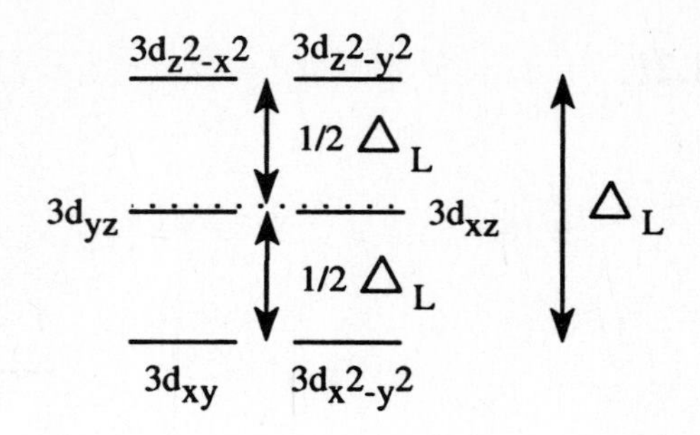

4.29.

$$E_{stabilization} = E_{destabilization}$$

$$2(0.19\Delta + x) + 2x = 0.79\Delta - x$$

$$0.38\Delta + 4x = 0.79\Delta - x$$
$$5x = 0.41\Delta$$
$$x = 0.082\Delta$$

0.79 Δ
0.79 Δ - x
x
0.19Δ
0.19 Δ + x

4.31. High- and low-spin tetrahedral complexes

d^3	e^3 and $e^2t_2^1$	d^5	$e^4t_2^1$ and $e^2t_2^3$
d^4	e^4 and $e^2t_2^2$	d^6	$e^4t_2^2$ and $e^3t_2^3$

4.33.

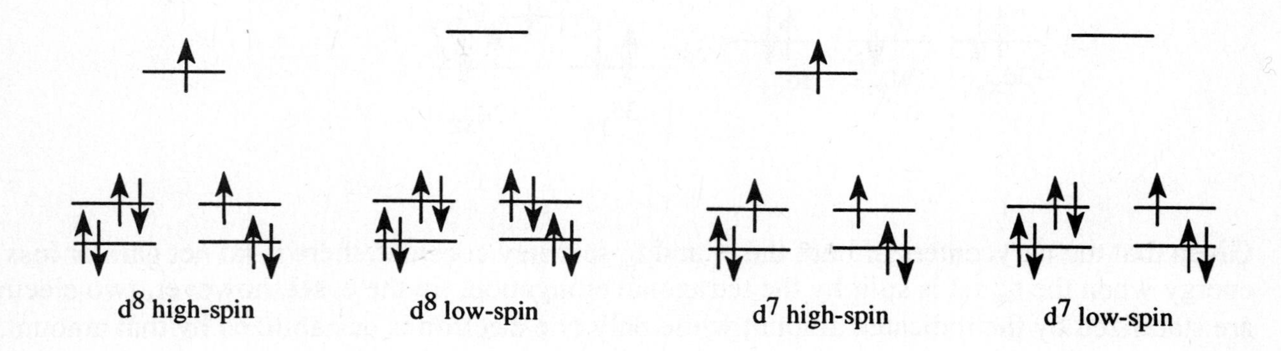

d^6 would not have high- and low-spin possibilities. There would always be two unpaired electrons in the middle degenerate set of d orbitals.

4.35.
(a) d^4, octahedral, low-spin (2)
(b) d^6, tetrahedral, high-spin (4)
(c) d^9, square planar (1)
(d) d^7, octahedral, high-spin (3)
(e) d^2, cubic (2)
(f) d^8, octahedral with tetragonal elongation (2)

4.37. d^5 high-spin $\quad 3(2/5\ \Delta_o) - 2(3/5\ \Delta_o) = 0$
low-spin $\quad 5(2/5\ \Delta_o) - 2P = 2\ \Delta_o - 2P$

d^6 high-spin $\quad 4(2/5\ \Delta_o) - 2(3/5\ \Delta_o) = 2/5\ \Delta_o$
low-spin $\quad 6(2/5\ \Delta_o) - 2P = 12/5\ \Delta_o - 2P$

d^7 high-spin $\quad 5(2/5\ \Delta_o) - 2(3/5\ \Delta_o) = 4/5\ \Delta_o$
low-spin $\quad 6(2/5\ \Delta_o) - 1(3/5\ \Delta_o) - P = 9/5\ \Delta_o - P$

4.39. (a) $[Fe(H_2O)_6]^{3+}$ $\quad Fe^{3+}$ $\quad d^5$ $\quad$ water is midway in the spectrochemical series while Fe^{3+} is fairly small with a large charge; therefore pick weak-field, high-spin case:

$$CFSE = 3(2/5\ \Delta_o) - 2(3/5\ \Delta_o) = 0$$

(b) $[PtCl_6]^{2-}$ $\quad Pt^{4+}$ $\quad d^6$ $\quad$ while chloride is low in the spectrochemical series, Pt^{4+} is large and highly charged, therefore pick strong-field, low-spin case:

$$CFSE = 6(2/5\ \Delta_o) - 2P = 12/5\ \Delta_o - 2P$$

(c) $[Cr(NH_3)_6]^{3+}$ $\quad Cr^{3+}$ $\quad d^3$ $\quad CFSE = 3(2/5\ \Delta_o) = 6/5\ \Delta_o$

4.41.

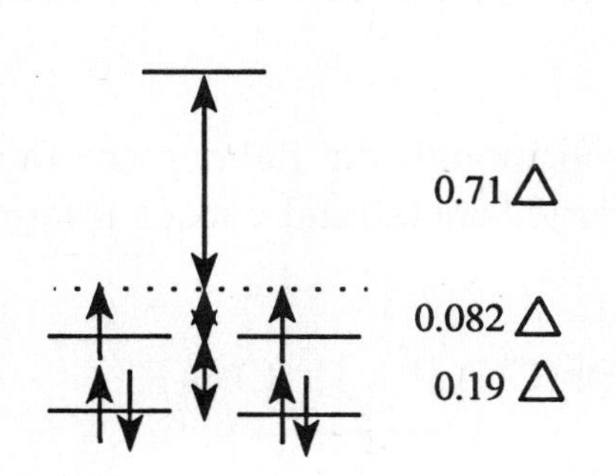

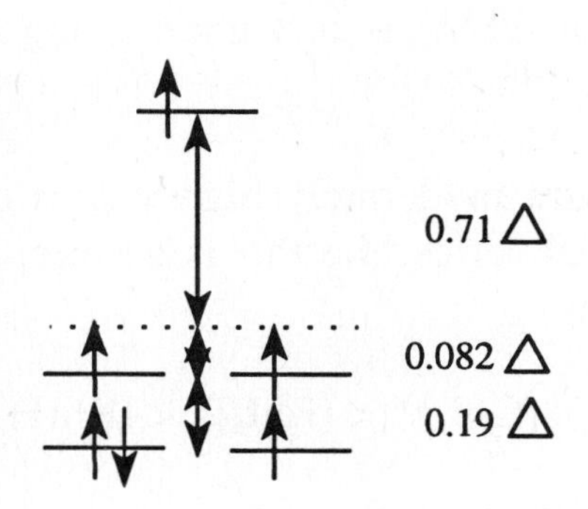

low-spin case:

$CFSE = 4(0.19\Delta + 0.082\Delta) + 2(0.082\Delta) - P$

$= 1.25\Delta - P$

high-spin case:

$CFSE = 3(0.19\Delta + 0.082\Delta)$
$+ 2(0.082\Delta) - 0.71\Delta$
$= 0.27\Delta$

The low-spin case will be favored as long as 0.98Δ is greater than P.

4.43.

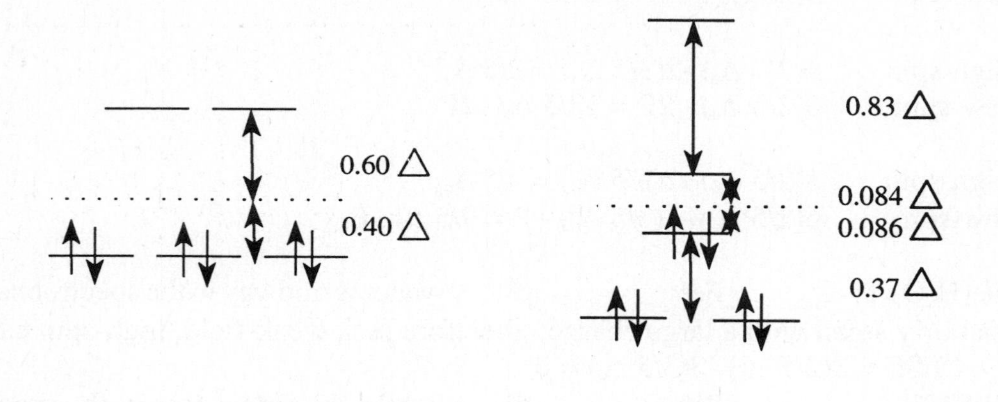

octahedral | square pyramidal

$$\text{CFSE} = 6(0.40\Delta) - 2P = 2.40\Delta - P$$

$$\text{CFSE} = 4(0.37\Delta + 0.086\Delta) + 2(0.086\Delta) = 2.00\Delta - P$$

Since 0.40Δ of CFSE is lost upon going to the square pyramidal complex, the substitution is probably slow.

4.45. These are moderately high values of Δ_o. Even though the halides are fairly low in the spectrochemical series, the Pt^{4+} is a large, highly charged metal and causes a large crystal field splitting energy.

4.47. $[CoI_6]^{4-} < [CoI_6]^{3-} < [RhI_6]^{3-} < [Rh(H_2O)_6]^{3+} < [Rh(CN)_6]^{3-} < [Ir(CN)_6]^{3-}$

4.49. $[Re(CN)_6]^{3-}$ contains Re^{3+} (d^4) in a strong-field, low-spin configuration and therefore should have two unpaired electrons (t_{2g}^4). $[MnCl_6]^{4-}$ contains Mn^{2+} (d^5) in a weak-field high-spin configuration and therefore should have five unpaired electrons ($t_{2g}^3e_g^2$).

4.51. Using only electrostatic arguments, the F^- ligand is a small charged ion and therefore should be able to get close to and effectively split the d orbitals of a transition metal atom or ion. The $P(C_6H_5)_3$ ligand, on the other hand, is neutral and much larger and therefore should not be able to get as close to or effectively split the d orbitals of a given metal. (It is π-interactions that make the fluoride a weak-field ligand and triphenyl phosphine strong-field.)

4.53. d^6 octahedral: strong-field: t_{2g}^6, $n = 0$, $\mu_S = 0$
weak-field: $t_{2g}^4e_g^2$, $n = 4$, $\mu_S = [(4)(4+2)]^{1/2} = 4.90$ BM
d^6 tetrahedral: strong-field: $e^4t_2^2$, $n = 2$, $\mu_S = [(2)(2+2)]^{1/2} = 2.83$ BM
weak-field: $e^3t_2^3$, $n = 4$, $\mu_S = 4.90$ BM

4.55.

	μ_{exptl} (BM)	# of unpaired electrons	M^{n+}	d^m	Elect. config.	Spin state
$[Mn(CN)_6]^{4-}$	1.8	1	Mn^{2+}	d^5	t_{2g}^5	low-spin
$[Mn(CN)_6]^{3-}$	3.2	2	Mn^{3+}	d^4	t_{2g}^4	low-spin
$[Mn(NCS)_6]^{4-}$	6.1	5	Mn^{2+}	d^5	$t_{2g}^3e_g^2$	high-spin
$[Mn(acac)_3]$	5.0	4	Mn^{3+}	d^4	$t_{2g}^3e_g^1$	high-spin

4.57. $[CoI_6]^{3-}$ Co^{3+} d^6 high-spin $t_{2g}^4e_g^2$ $n = 4$ $\mu_S = [(4)(6)]^{1/2} = 4.90$ BM

4.59. For the $[RuF_6]^{4-}$ ion, $\mu = 2.84\,(\chi_M T)^{1/2} = 2.84[(1.01 \times 10^{-2})(298)]^{1/2} = 4.93$ BM which is consistent with 4 unpaired electrons. For the $[Ru(PR_3)_6]^{2+}$ ion, $\chi_M = 0$ indicates that $\mu = 0$ and therefore there are no unpaired electrons. Four unpaired electrons is consistent with a d^6 ion in a weak field while zero unpaired electrons corresponds to a d^6 ion in a strong field.

4.61. Yes, it would be surprising if all the compounds involving these complex ions were to be the same color. Given the spectrum (pun intended) of crystal field splitting abilities of the ligands involved, these complexes should absorb at a variety of wavelengths and therefore be of various colors.

4.63. The octahedral $[Ni(H_2O)_6]^{2+}$ has a $t_{2g}^6e_g^2$ configuration corresponding to two unpaired electrons. The $[Ni(CN)_4]^{2-}$ must be a strong-field, square planar complex in order to be diamagnetic. It must absorb at considerably higher frequencies (most likely into the ultraviolet) than the hexaaquo complex and therefore transmits or reflects all the visible frequencies and appears colorless.

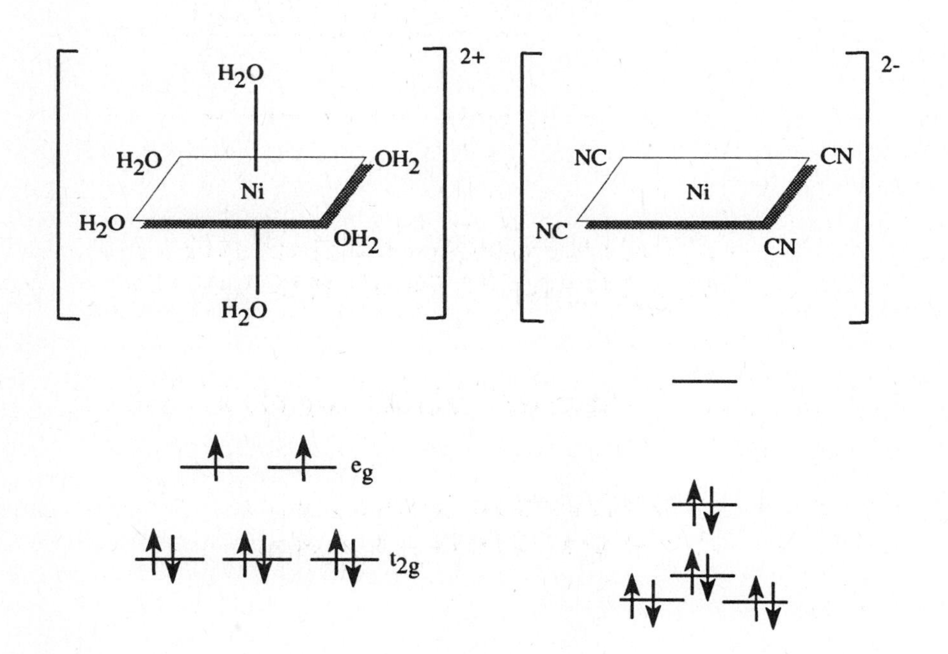

4.65. To picture the bonding in coordination compounds as strictly ionic is, at first thought, rather unrealistic. However, this theory does a surprisingly good job at accounting for the existence and high stability of these compounds as well as their magnetic characteristics and colors. The experimental evidence for the theory, as covered in this chapter, comes from two major types of studies: magnetic measurements and absorption spectroscopy. Magnetic measurements generate values for the molar magnetic susceptibilities of a series of coordination compounds. Molar susceptibilities, in turn, generate molecular-level magnetic moments that can be related to the number of unpaired electrons and electronic configurations of a complex. These results are explained quite adequately by the crystal field theory. Absorption spectroscopy gives information on the frequencies of light absorbed by coordination compounds. These frequencies vary in ways (depending on the size and charge of the metal and ligands) that can be explained in many cases by the ionic crystal field theory.

Chapter 5
Rates and Mechanisms of Reactions of Coordination Compounds

The sections and subsections in this chapter are listed below.

5.1 A Brief Survey of Reaction Types
5.2 Labile and Inert Coordination Compounds
5.3 Substitution Reactions of Octahedral Complexes
- Possible Mechanisms
- Experimental Complications
- Evidence for Dissociative Mechanisms
- Explanation of Inert versus Labile Complexes

5.4 Redox or Electron Transfer Reactions
- Outer-Sphere Mechanisms
- Inner-Sphere Mechanisms

5.5 Substitution Reactions in Square Planar Complexes: The Kinetic Trans Effect

Chapter Objectives

You should be able to

- recognize the various types of reactions that coordination compounds undergo
  - substitution (including hydrolysis and anation),
  - dissociation,
  - addition,
  - electron transfer (including oxidative-addition and reductive-elimination),
  - reactions of coordinated ligands
- write expressions for overall and stepwise equilibrium constants for a given substitution reaction
- distinquish between and correctly use the kinetic terms labile and inert as well as the thermodynamic terms stable and unstable
- write dissociative (D), associative (A), and interchange (I) mechanisms for substitution reactions
- explain, when determining the rate law of a given reaction, how the concentration of a reactant may be masked by certain experimental conditions
- cite evidence from rates of exchange of water molecules, as well as anation and aquation reactions, that substitution reactions of octahedral complexes most often follow a dissociative mechanism
- discuss the contributions of metal size and charge, steric hindrance among ligands, overall charge on the complex, and M-L covalent overlap to the energy of activation of the bond-breaking, rate-determining step of a substitution reaction
- discuss the contribution of the change of crystal field stabilization energy to the energy of activation of the bond-breaking, rate-determining step of a substitution reaction
- summarize why complexes of the first row transition metal ions, with the exception of Cr^{3+} and Co^{3+}, are generally labile while those of most second and third row transition metal ions are inert.
- describe outer-sphere electron transfer mechanisms and how their rates are a function of the relative M-L distances in the reactants
- describe inner-sphere electron transfer mechanisms and how their rates are a function of the polarizability of the bridging ligand
- describe the kinetic trans effect and rationalize it in terms of the polarizability of the *trans*-directing ligand
- be able to use the kinetic trans effect to outline the synthesis of various square planar complexes

Solutions to Odd-Numbered Problems

5.1. $$K_1K_2K_3K_4 = \frac{[\{Cu(NH_3)(H_2O)_3\}^{2+}]}{[\{Cu(H_2O)_4\}^{2+}][NH_3]} \times \frac{[\{Cu(NH_3)_2(H_2O)_2\}^{2+}]}{[\{Cu(NH_3)(H_2O)_3\}^{2+}][NH_3]} \times$$

$$\frac{[\{Cu(NH_3)_3(H_2O)\}^{2+}]}{[\{Cu(NH_3)_2(H_2O)_2\}^{2+}][NH_3]} \times \frac{[\{Cu(NH_3)_4\}^{2+}]}{[\{Cu(NH_3)_3(H_2O)\}^{2+}][NH_3]}$$

$$= \frac{[\{Cu(NH_3)_4\}^{2+}]}{[\{Cu(H_2O)_4\}^{2+}][NH_3]^4} = \beta_4$$

5.3. $$K_1 = \frac{[Cr(CO)_5PPh_3]}{[Cr(CO)_6][PPh_3]}$$

$$K_2 = \frac{[Cr(CO)_4(PPh_3)_2]}{[Cr(CO)_5PPh_3][PPh_3]} \qquad \beta_3 = K_1K_2K_3 = \frac{[Cr(CO)_3(PPh_3)_3]}{[Cr(CO)_6][PPh_3]^3}$$

$$K_3 = \frac{[Cr(CO)_3(PPh_3)_3]}{[Cr(CO)_4(PPh_3)_2][PPh_3]}$$

5.5. The concentration of water would not be found in the mass action expression representing the reaction of dilute aqueous acetic acid with water. $[H_2O]$ would be assumed to be a constant in this case and would be incorporated into the value of K_a of the acid.

5.7. (a) dissociation
(b) addition
(c) substitution
(d) oxidation and addition, i.e., oxidative-addition
(e) redox and substitution

5.9. The O^{17} label would end up in the water molecule which is given as a product of the reaction. The Cr^{3+}-O bonds in the hexaaquochromium(III) ion are not broken in this reaction of a coordinated ligand. Instead, the OH^- reactant abstracts a proton (H^+) from one of the waters leaving an OH^- ligand in its place in the coordination sphere.

5.11. Strong-field, low-spin d^6 cobalt(III) complexes are inert to substitution because they lose crystal field stabilization energy upon breaking a metal-ligand bond to form a five-coordinate transition state. Accordingly, it is often easier (kinetically) to prepare cobalt(III) compounds by oxidizing Co(II) complexes which are not inert.

5.13.

(a)	$[Co(NH_3)_6]^{2+}$	Co^{2+}	labile
(b)	$[Co(NH_3)_5NO_2]^{2+}$	Co^{3+}, d^6, moderately strong-field	inert
(c)	$[CoI_6]^{3-}$	Co^{3+}, d^6, weak-field	labile
(d)	$[Fe(H_2O)_5(NCS)]^{2+}$	Fe^{3+}	labile
(e)	$[Ni(en)_3]^{2+}$	Ni^{2+}	labile
(f)	$[IrCl_6]^{3-}$	Ir^{3+}	inert

5.15. $[Ni(CN)_4]^{2-}$ | Ni^{2+} | d^8 | square planar, room for attack of incoming ligand; assuming an associative mechanism — most likely to be labile

$[Mn(CN)_6]^{3-}$ | Mn^{3+} | d^4 | in between, high charge density but only a moderate loss CFSE — an in between case

$[Cr(CN)_6]^{3-}$ | Cr^{3+} | d^3 | high charge density; significant loss of CFSE— most likely to be inert

5.17. The situation we are faced with is as follows:

$$(1)\ ML_5X \underset{k_{-1}}{\overset{k_1}{\rightleftharpoons}} ML_5 + X \qquad \text{(fast)}$$

$$(2)\ ML_5 + Y \xrightarrow{k_2} ML_5Y \qquad \text{(slow)}$$

The rate-determining step is (2), so therefore,

$$\text{Overall rate} = \text{Rate of step (2)} = k_2[ML_5][Y]$$

Now the fast step (1) will establish an equilibrium "behind" the rate-determining second step such that
$$k_1[ML_5X] = k_{-1}[ML_5][X]$$

Solving this last equilibrium expression for $[ML_5]$ yields $[ML_5] = k_1[ML_5X] / k_{-1}[X]$

Substituting this expression into the one for the overall rate yields

$$\text{Overall rate} = \frac{k_2k_1[ML_5X]\,[Y]}{k_{-1}[X]}$$

5.19. Since the first step is now considered to be faster than the second, the concentration of ML_5XY would now be able to build up to a significant extent. Given that the second step is now much slower than the first, not much of the ML_5XY would be used up in the rate-determining step. It follows that we expect the concentration of ML_5XY to be relatively high.

5.21. In the absence of experimental complications, an associative mechanism should yield a rate law first order in both the reactant complex and the entering ligand while the dissociative mechanism would yield a rate law first order only in the reactant complex. In addition, an A mechanism might be proven by the isolation of the seven-coordinate intermediate while the D could be verified by the isolation of the five-coordinate intermediate.

5.23. Dissociative or interchange-dissociative mechanisms for the substitution reactions in octahedral coordination compounds are supported by the observed rates of (a) water exchange, (b) anation, and (c) aquation. The rates of exchange of water molecules in a hydrated metal ion is observed to be faster for metals of lower charge and/or larger size. Specifically, as the charge density decreases, the rate of exchange is found to be faster. This observation is consistent with a dissociative (or interchange-dissociative) mechanism because decreasing the charge density weakens the bond between the metal and the ligand-to-be-replaced and therefore leads to a faster rate-determining dissociative step. The fact that the rates of anation reactions are independent of the incoming anion is also consistent with the rate-determining step being the splitting off of a coordinated ligand followed by the addition of the incoming anion. Finally, the fact that the rates of aquation reactions *are* found to be dependent on the identity of the ligand to be replaced is also consistent with a dissociative mechanism. Data from all three of these types of reactions indicate that the strength of the original metal-ligand bond (broken in the rate-determining step of dissociative mechanism) determines the rate at which this ligand is replaced.

5.25. These data are generally consistent with a dissociative mechanism. Log k does not vary appreciably with the identity of the entering ligand, L. (The variations that do occur might very well be due to the ligand L "lurking" outside the coordination sphere of the reactant as in an I_d mechanism.) The rate-determining step would be

$$[Cr(NH_3)_5H_2O]^{3+} \xrightarrow{\text{slow}} [Cr(NH_3)_5]^{3+} + H_2O$$

Since the rate-determining step does not involve the L ligand, the rates are not dependent on [L].

5.27. If the rate constant is $1.0 \times 10^{-4}\ s^{-1}$ then log k is -4.0, a value of the same order of magnitude as those for the anation reactions of Problem 5.26. This similarity implies that again the rate-determining step is the breaking of the Co^{3+}-OH_2 bond and this reaction occurs via a dissociative mechanism.

5.29. If the rate constant is $3 \times 10^{-6}\ M^{-1}s^{-1}$ then log k is -5.5, a value that differs significantly from most of the other data given in Problem 5.28. This data seems to lead to the conclusion that these reactions occur by an associative rather than a dissociative mechanism. If the splitting of a Cr^{3+}-OH_2 were the rate-determining step, than the rates of all these reactions should all be of the same order of magnitude.

5.31. The table shows that these aquation reactions depend quite strongly on the identity of the L^- ligand. This result is consistent with a rate-determining step in which the M-L bond is broken to form a five-coordinate transition state.

5.33. The change in the CFSE on going from the octahedral reactant to the five-coordinate intermediate directly affects the kinetics of substitution reactions. If CFSE is gained by forming the five-coordinate intermediate, the energy of activation (E_a) will be reduced by that amount and the reaction will be faster than expected in the absence of crystal field effects. If, on the other hand, CFSE is lost upon forming the intermediate, the E_a will be increased by that amount and the reaction will be slower.

5.35. For a d^6 octahedral case, CFSE = $6(0.40\Delta_o) - 2P = 2.40 - 2P$
For the square pyramidal case, CFSE = $4(0.457\Delta_o) + 2(0.086\Delta_o) = 2.00 - 2P$
ΔCFSE = $-0.40\Delta_o$

5.37. For a d^8 octahedral case, CFSE = $6(0.40\Delta_o) - 2(0.60\Delta_o) = 1.20\Delta_o$
For a d^8 trigonal bipyramidal case, CFSE = $4(0.272\Delta_o) + 3(0.082\Delta_o) - 0.707\Delta_o = 0.627\Delta_o$
(a) There is a substantial loss ($1.20\Delta_o - 0.627\Delta_o = -0.57\Delta_o$) of CFSE upon going to the trigonal bipyramidal intermediate transition state, therefore we expect this reaction to be slower than it would be in the absence of crystal field effects. On the other hand, these ions only have a +2 charge and therefore would probably not be classified as inert.
(b) The loss in CFSE for the square pyramidal state (see Table 5.4) is only $0.20\Delta_o$. Since loss of CFSE adds directly to the energy of activation, the E_a will most likely be smaller for the mechanism involving

the square pyramidal transition state. This mechanism will therefore be favored rather than the one involving the trigonal bipyramidal intermediate.

5.39. Ni^{2+} ions have only a +2 charge and therefore do not have a large enough charge density to be classified as inert. In order for a first row transition metal to be inert it appears that it must be +3 charged and lose $0.20\Delta_o$ or more of CFSE.

5.41. This cross-reaction most likely proceeds by an outer-sphere mechanism. It is particularly fast because there is little change in the M-L distances in either the iron or iridium complexes. (Recall that the $Fe^{2+/3+}$-CN distances vary by only 0.03Å [see Equation (5.42)]. $Ir^{3+/4+}$-Cl distances most likely do not change very much either. They, like those involving $Fe^{2+/3+}$, involve low-spin d^5 and d^6 complexes which do not have electrons occupying e_g orbitals that point directly at ligands.)

5.43. When the Co^{3+} in the $[Co(en)_3]^{3+}$ complex accepts an electron, it must be placed in an e_g orbital that points directly at the ligands. Therefore, the Co^{2+}-N distance in $[Co(en)_3]^{2+}$ will be appreciably longer than in $[Co(en)_3]^{3+}$. In order for the electron transfer to take place, we picture the M-L distances being adjusted to an intermediate position. This process will take a considerable amount of energy that will be added to the energy of activation for the rate-determining step and consequently make this electron transfer rather slow.

5.45. The rate-determining step is most likely the second step of the mechanism, that is, the readjustment of the bond distances within the ion pair followed by the rapid transfer of the electron from one metal to the other. Using the example detailed in Equations (5.38) to (5.40), the rate of the rate-determining step would be

$$\text{Rate} = k[\{Ru(H_2O^{17})_6\}^{3+} / \{Ru(H_2O)_6\}^{2+}]$$

The slower rate-determining second step would force an equilibrium situation to be set up in the first step in which the ion pair is formed. That is,

$$k_f[\{Ru(H_2O^{17})_6\}^{3+}][\{Ru(H_2O)_6\}^{2+}] = k_r\,[\{Ru(H_2O^{17})_6\}^{3+} / \{Ru(H_2O)_6\}^{2+}],$$

where k_f and k_r are the rate constants in the forward and reverse directions, respectively.

Solving this last equation for the concentration of the ion pair and substituting that into the expression for the rate-determining step yields the following second-order rate law:

$$\text{Rate} = (kk_f/k_r)[\{Ru(H_2O^{17})_6\}^{3+}][\{Ru(H_2O)_6\}^{2+}]$$

5.47. Other contributions to the energy of activation might include (a) the positive contribution to the potential energy when similarly-charged coordination spheres come together to form the ion pair and (b) the rearrangement of the solvent structure (the solvent in the cases we have considered has been water) around the newly reconfigured coordination spheres when the M-L distances are adjusted and the electron transferred from one metal to another.

5.49. $[Co(NH_3)_5SCN]^{2+}$ + $[Cr(H_2O)_6]^{2+}$ → $[Cr(H_2O)_5NCS]^{2+}$ + $[Co(H_2O)_6]^{2+}$ + $5NH_3$

Inner-Sphere Electron Transfer Mechanism:

$[Cr(H_2O)_6]^{2+}$ → $[Cr(H_2O)_5]^{2+}$ + H_2O

$[Co(NH_3)_5SCN]^{2}$ + $[Cr(H_2O)_5]^{2+}$ ⇆ $[(NH_3)_5Co^{III}—SCN^-—Cr^{II}(H_2O)_5]^{4+}$

$[(NH_3)_5Co^{III}—SCN^-—Cr^{II}(H_2O)_5]^{4+}$ → $[(NH_3)_5Co^{II}—SCN^-—Cr^{III}(H_2O)_5]^{4+}$

e^-

$[(NH_3)_5Co^{II}—SCN^-—Cr^{III}(H_2O)_5]^{4+}$ → $[Co(NH_3)_5]^{2+}$ + $[Cr(H_2O)_5NCS]^{2+}$

$[Co(NH_3)_5]^{2+}$ —xs H_2O→ $[Co(H_2O)_6]^{2+}$ + $5NH_3$

labile

5.51. Mechanism:

$[Cr^*(H_2O)_6]^{2+}$ —slow→ $[Cr^*(H_2O)_5]^{2+}$ + H_2O

labile

$[Cr^*(H_2O)_5]^{2+}$ + $[Cr(H_2O)_5X]^{2+}$ ⇆ $[(H_2O)_5Cr^{III}\text{-}X\text{-}{}^*Cr^{II}(H_2O)_5]^{4+}$

inert

$[(H_2O)_5Cr^{III}\text{-}X\text{-}{}^*Cr^{II}(H_2O)_5]^{4+}$ → $[(H_2O)_5Cr^{II}\text{-}X\text{-}{}^*Cr^{III}(H_2O)_5]^{4+}$

e^-

$[(H_2O)_5Cr^{II}\text{-}X\text{-}{}^*Cr^{III}(H_2O)_5]^{4+}$ → $[Cr(H_2O)_5]^{2+}$ + $[Cr^*(H_2O)_5X]^{2+}$

$[Cr(H_2O)_5]^{2+}$ + H_2O → $[Cr(H_2O)_6]^{2+}$

The rates of these reactions increase in the order X^- = F^- to Cl^- to Br^- due to the increase in the polarizability of the bridging ligand. In the bridged complex, the Cr^{3+} polarizes X^-. The induced dipole in X draws electron density away from the Cr^{2+} and serves as a conduit for the transfer of the electron from Cr^{2+} to the Cr^{3+}. As the polarizability of X^- increases, its ability to facilitate the electron transfer increases and the rates of these reactions increase.

5.53. The very last step in the bottom right-hand corner is controversial. Ordinarily, the chloride ligand is the most readily replaced. However, in this case, the greater *trans*-directing effect of the chloride allows the pyridine trans to it to be preferentially replaced. (The charges on the various complexes are omitted for clarity.)

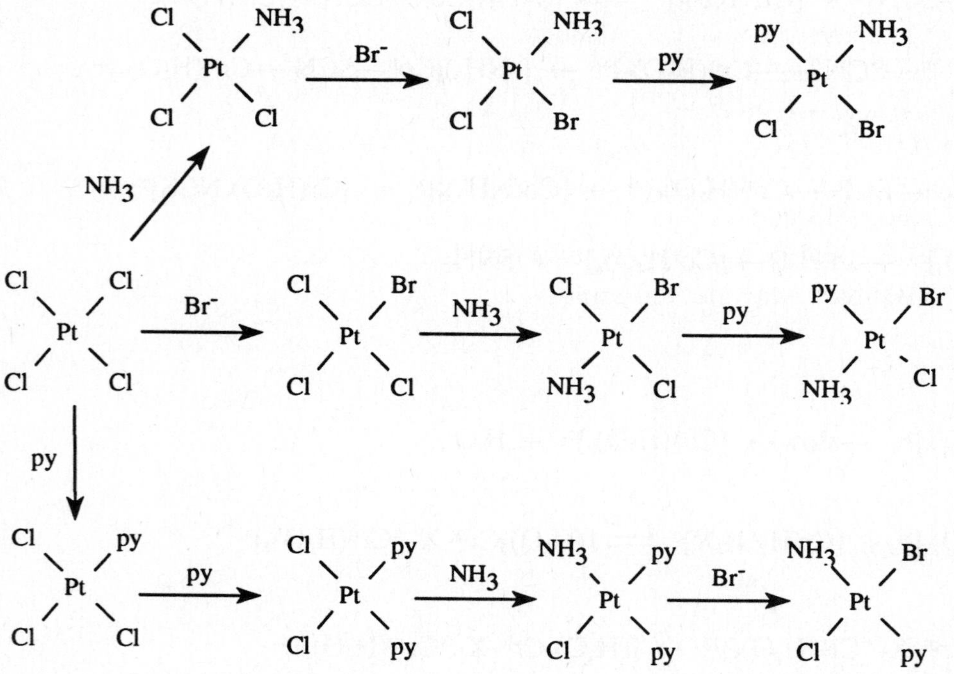

5.55. Given the fact that the OH^- is a very small, difficult-to-polarize ligand, one would speculate that it would be very low, perhaps the lowest of those given in the *trans*-directing series.

5.57. (a) carbonyl-*trans*-dichloro(pyridine)platinum(II)
(b) *cis*-diamminechloronitroplatinum(II)
(c) aquo-*trans*-dichlorothiocyanatoplatinate(II)
(d) ammine-*trans*-dichlorocyanoplatinate(II)

Chapter 6
Applications of Coordination Compounds

The sections and subsections in this chapter are listed below.

6.1 Applications of Monodentate Complexes
6.2 Two Keys to the Stability of Transition Metal Complexes
  Hard and Soft Acids and Bases
  The Chelate Effect
6.3 Applications of Multidentate Complexes
6.4 Chelating Agents as Detergent Builders
6.5 Bioinorganic Applications of Coordination Chemistry
  Hemoglobin
  Therapeutic Chelating Agents for Heavy Metals
  Platinum Antitumor Agents

Chapter Objectives

You should be able to

- cite examples of transition metal complexes involving monodentate ligands in qualitative analysis, dyes, silver and gold ore processing, nickel purification, and black/white photography
- define, characterize, rationalize, and use the concept of hard and soft acids and bases as applied to the stability of metal-ligand interactions
- define, explain, and give examples of the chelate effect
- cite and explain applications of multidentate complexes drawn from complexometric quantitative analytical methods
- explain why EDTA is used to remove hard water deposits from hot water boilers and heaters and is often added to foods and other consumer products
- explain the function of detergent builders and the advantages and disadvantages of using phosphates, nitrilotriacetic acid, and carbonates to carry out that function over the years
- explain the role of hemoglobin and its oxygen complexes in the process of respiration
- explain how carbon monoxide and cyanide poisoning work
- explain how and why EDTA, pencillamine, and British anti-lewisite function as therapeutic chelating agents for heavy metals
- briefly cite some of the history and symptoms of lead and mercury poisoning
- explain how cisplatin and its derivatives serve as antitumor agents

6.1. Diamminesilver(I), $Ag(NH_3)_2^+$, contains Ag(I), a d^{10} metal. With no vacancies in d orbitals, coordination compounds containing this cation cannot readily absorb visible light and so remain colorless. On the other hand, tetraamminecopper(II), $Cu(NH_3)_4^{2+}$, contains Cu(II), a d^9 metal that does have vacancies in its d orbitals and therefore readily absorbs visible light.

6.3. Cyanide acts as a halide. In fact, ions such as cyanide (CN^-), thiocyanate (SCN^-), and others (see Chapter 18 for more details) are often referred to as pseudohalides because of their similarities to these ions. Accordingly, cyanide can often be quantitatively analyzed in the same way as the halide ions. One method of doing this is to add a slight excess of a solution of silver nitrate, $AgNO_3$, to the cyanide until silver cyanide, AgCN(s), is precipitated, isolated, and then weighed quantitatively. The Liebig method actually refers to the method in which a standardized solution of silver nitrate is used to titrate a solution of cyanide until the first appearance of a precipitate of silver dicyanoargentate(I), $Ag[Ag(CN)_2]$. This reaction becomes the basis of a titrimetric determination of cyanide.

$$2CN^-(aq) + 2Ag^+ \rightarrow Ag[Ag(CN)_2](s)$$

6.5. According to Equation (6.7), the reaction (reproduced below) shifts back to the left when the temperature is raised.

$$Ni(s) + 4CO(g) \rightleftharpoons Ni(CO)_4(g)$$

In order to force the equilibrium back to the left, heat must be present as a product. That is, the reaction must be exothermic.

6.7. Since soft acids are large, highly polarizable metals of low (even zero) oxidation states and soft bases are large, polarizable ligands of low charge, it follows that the soft-soft M-L bond between such species would be characterized by high London dispersion forces. That is, both the ligand and metal have electron clouds in which instantaneous dipoles could readily and often occur. Such dipoles could then induce dipoles in the other electron clouds. Such a situation is characteristic of interactions with substantial London forces. On the other hand, since hard acids are small, often highly charged metal ions (Lewis acids) with large charge densities and low polarizabilities whereas hard bases are small, highly electronegative ligands (Lewis bases) of low polarizabilities, these hard-hard M-L interactions would not be characterized by high London forces.

6.9. The diacetatotetraaquoiron(II) ion involves two acetate ions acting as monodentate ligands. The malonate ion has two acetate moities bound together through a common $-CH_2-$ group. Both of the acetate moities can act as Lewis bases and make malonate a bidentate ligand. The difference between the equilibrium constants, then, is an example of the chelate effect. The malonate complex involves a chelating, multidentate ligand that is more stable than the equivalent compound involving monodentate ligands. These two situations are shown in the following two equations. (The acetates could also form a trans complex.)

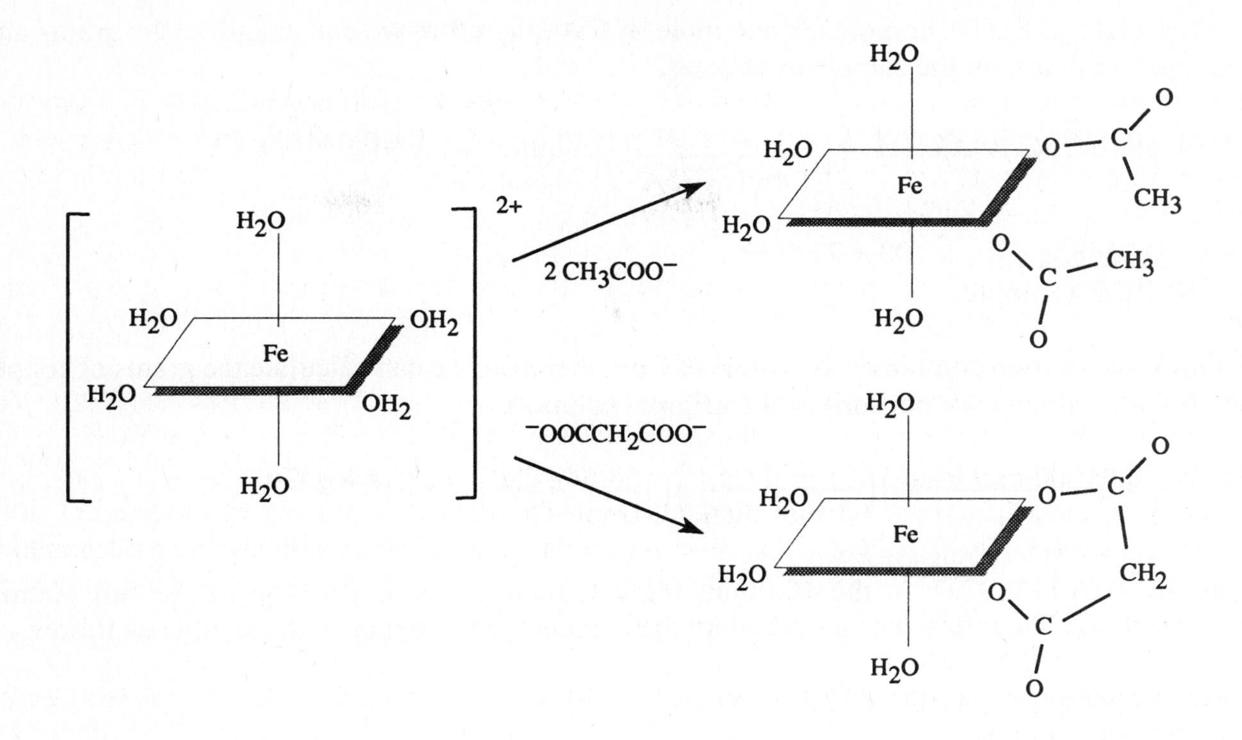

6.11. Acetone is a monodentate ligand whereas acetylacetonate, $[CH_3COCHCOCH_3]^-$, made up of two acetone [or acetyl, CH_3CO] moities connected together by a common -CH- group, is bidentate. The difference between the stability of acetone and acetylacetonate, then, is due to the chelate effect. The acac complexes involve chelating, bidentate ligands that form more stable compounds than the equivalent compounds involving two monodentate ligands.

6.13. Using Equations (6.9) and (6.10) as shown below, we can calculate the equilibrium constant for the substitution of three ethylenediamine ligands for six ammines in the Ni(II) coordination sphere.

$$[Ni(NH_3)_6]^{2+}(aq) \rightarrow Ni^{2+}(aq) + 6NH_3(aq) \qquad K = 1/\beta = 1/(4.0 \times 10^8)$$
$$Ni^{2+}(aq) + 3en(aq) \rightarrow [Ni(en)_3]^{2+}(aq) \qquad \beta = 2.0 \times 10^{18}$$
$$[Ni(NH_3)_6]^{2+}(aq) + 3en(aq) \rightarrow [Ni(en)_3]^{2+}(aq) + 6NH_3(aq) \qquad K = \frac{2.0 \times 10^{18}}{4.0 \times 10^8} = 5.0 \times 10^9$$

6.15. The dmgH ligand has a molecular formula of $C_4H_7O_2N_2$ and a molecular weight of 115.11 g/mol. Since one mole of Ni yields one mole of $Ni(dmgH)_2$, we can calculate the mass of nickel in the ore as follows:

$$0.7815g\ Ni(dmgH)_2 \left(\frac{1\ mol\ Ni(dmgH)_2}{288.91g\ Ni(dmgH)_2}\right)\left(\frac{1\ mol\ Ni}{1\ mol\ Ni(dmgH)_2}\right)\left(\frac{58.69\ g\ Ni}{1mol\ Ni}\right) = 0.1588g\ Ni$$

$$\%\ Ni = \frac{0.1588g\ Ni}{0.3456g\ ore} \times 100 = 45.94\%$$

6.17. One mole of EDTA complexes one mole of Ca^{2+}, therefore we can calculate the grams and percentage of calcium in the sample as follows.

$$0.02394L\left(\frac{0.04672 \text{ mol EDTA}}{L}\right)\left(\frac{1\text{mol Ca}}{1 \text{ mol EDTA}}\right)\left(\frac{40.08 \text{ g Ca}}{1\text{mol Ca}}\right) = 0.04483\text{g Ca}$$

$$\% \text{ Ca} = \frac{0.04483\text{g Ca}}{0.2000 \text{ g sample}} \times 100 = 22.41\%$$

6.19. One mole of trien complexes one mole of Cu^{2+}, therefore we can calculate the grams of copper in the 10.00mL aliquot (or portion) of the original solution.

$$0.02275L\left(\frac{0.01000 \text{ mol trien}}{L}\right)\left(\frac{1 \text{ mol Cu}}{1 \text{ mol trien}}\right)\left(\frac{63.54\text{g Cu}}{1\text{mole Cu}}\right) = 0.01446\text{g Cu}$$

If there are 0.01446g Cu^{2+} in the 10.00mL aliquot, there must be 0.1446g in the full 100mL volumetric flask. Therefore we can calculate the percentage of copper in the sample as follows.

$$\% \text{ Cu} = \frac{0.1446\text{g Cu}}{0.2005\text{g sample}} \times 100 = 72.10\%$$

$$6.21. \quad \left(\frac{8.7\text{g P}}{100\text{g sample}}\right)\left(\frac{1 \text{ mol P}}{30.97\text{g P}}\right)\left(\frac{1\text{mol } Na_5P_3O_{10}}{3\text{mol P}}\right)\left(\frac{367.85\text{g } Na_5P_3O_{10}}{1\text{mol } Na_5P_3O_{10}}\right) = \frac{34\text{g } Na_5P_3O_{10}}{100\text{g sample}} = 34\%$$

6.23. Dear Econ Major:

Phosphates have been an important component of synthetic detergents because they let the soaplike part of a detergent do its job unhindered by hard water. In other words, phosphates render the hard water ions (calcium, magnesium, and iron) incapable of interfering with the action of the soaplike molecules in detergents. Phosphates have also been important because they are inexpensive and readily available to the detergent manufacturer. (See footnote at the end of this chapter.)

6.25. Dear Biology Major:

You may recall from your mind-bending introductory chemistry class that things are colored because they absorb some wavelengths of visible light and either allow other wavelengths through or reflect them to our eyes so that we perceive something to be colored. Sometimes this process is expressed by a colorwheel like the one shown below. This wheel says that if an object, for example, absorbs blue-green wavelengths it will therefore reflect or transmit red light to our eyes. Red light has a wavelength in the range of 750-610nm. What causes light to be absorbed? Usually this is attributed to various energy levels in an atom or molecule. In the hemoglobin of the blood there are a number of iron ions surrounded by a so-called heme group. This heme group causes a split among the d orbitals of the iron resulting in the absorption of blue-green light so that our blood appears red. When the blood is oxygenated, an oxygen molecule (O_2) also exerts a slight influence on the split among the d orbitals of the iron. Now the "oxyhemoglobin" absorbs at a slightly different range of wavelengths so that the blood now appears a brighter shade of red. (See footnote at the end of this chapter.)

Wavelength Range (nm)	Color
750-610	red
610-590	orange
590-570	yellow
570-500	green
500-450	blue
450-370	violet

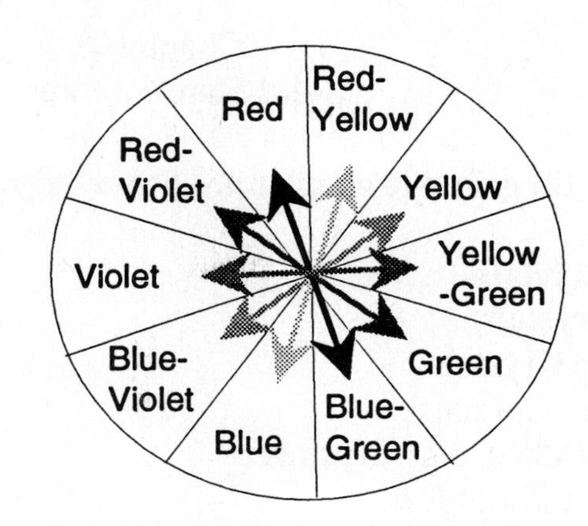

6.27. A low-spin d^6 Fe^{2+} has a t_{2g}^6 configuration in which all the electrons of the metal occupy orbitals that point in between the ligands of a square planar coordination sphere. When the Fe^{2+} changes to a high-spin state with a $t_{2g}^4e_g^2$ electronic configuration, two metal d electrons now occupy orbitals that point directly at the ligands. The resulting extra repulsions between the metal and the ligands in the high-spin case cause the metal to be *effectively* larger (the ligands are pushed away somewhat) and make the metal ion less able to fit into the square planar site.

6.29. Lead (Pb^{2+}) is a soft metal cation. It prefers to interact with soft anions such as sulfide (S^{2-}) rather than a relatively hard anion like oxide (O^{2-}).

6.31. Of EDTA, penicillamine, and BAL, the latter two are optically active. Both have a carbon atom with four different groups attached to it. Such a "center of chirality" produces a lack of an internal mirror plane and optical activity.

6.33. Cysteine, with its soft mercaptan (-SH) group would be the better antidote for the soft heavy metals.

6.35. Cysteine is probably the better antidote against heavy metals because it forms 5-membered rings with metals whereas methionine forms 6-membered rings. Five-membered rings are generally more stable.

Footnote:
(Dear readers: I know you can do better than I did in writing letters to various academic majors and professionals. I would love to see the various short paragraphs that you all come up with. You'll be famous; I might put your paragraph in the next version of this Solution Manual! (If there is a next version.) Submit your entries to the author at Department of Chemistry, Allegheny College, Meadville, PA 16335.

Chapter 7
Solid State Structures

The sections and subsections in this chapter are listed below.

- 7.1 Types of Crystals
  - Ionic Crystals
  - Metallic Crystals
  - Covalent Crystals
  - Atomic/Molecular Crystals
- 7.2 A-type Crystal Lattices
  - A-type Lattices
- 7.3 AB_n-type Crystal Lattices
  - Cubic, Octahedral, and Tetrahedral Holes
  - Radius Ratios
  - Ionic Radii
  - AB Structures
  - AB_2 Structures
- 7.4 Structures Involving Polyatomic Molecules and Ions
- 7.5 Defect Structures
- 7.6 Spinel Structures: Connecting Crystal Field Effects with Solid State Structures

Chapter Objectives

You should be able to

- distinquish among and characterize the various types of crystals: ionic, metallic, covalent, and atomic/molecular
- identify and characterize common A-type structures by the coordination number of a given sphere, the number of spheres per unit cell, the fraction of space occupied, and the density expression
- identify and appreciate the significance of the Bravais lattices
- characterize atomic, van der Waals, and metallic radii
- calculate the density of a given metallic crystal knowing its structure and relevant atomic weight and radii data
- identify and determine the relative sizes of cubic, octahedral, and tetrahedral holes in various A-type structures
- use radius ratios to predict the type of hole occupied by the smaller species in an AB_n structure
- rationalize why the radius ratio represents the lower limit to the range in which a given type of hole will be occupied
- discuss the origin and variation of Shannon-Prewitt ionic radii
- identify and characterize common AB structures (rock salt, zinc blende, wurtzite, and cesium chloride) by the structure and coordination numbers of both their anions and cations
- calculate the density of a given compound knowing the type of AB structure it assumes
- identify and characterize common AB_2 structures (fluorite, cadmium iodide, rutile, and anti-fluorite) by the structure and coordination numbers of both their anions and cations
- identify and discuss AB_n structures involving polyatomic molecules and ions
- identify and characterize common defect structures such as Schottky and Frenkel defects and edge dislocations
- characterize normal and inverse spinel structures and provide a rationale for the structure of a given compound based on calculations of crystal field stabilization energies

7.1. Metallic Crystals -- a lattice of cations held together by a sea of free electrons, for example, metallic copper. Covalent Crystals -- atoms or groups of atoms in a lattice held together by an interlocking network of covalent bonds, for example, diamond. Atomic/Molecular Crystals -- a lattice of atoms or molecules held together by intermolecular forces (Van der Waals, dipole-dipole, hydrogen bonds), for example, argon, water.

7.3. (a) London dispersion forces -- temporary-dipole, induced-dipole forces

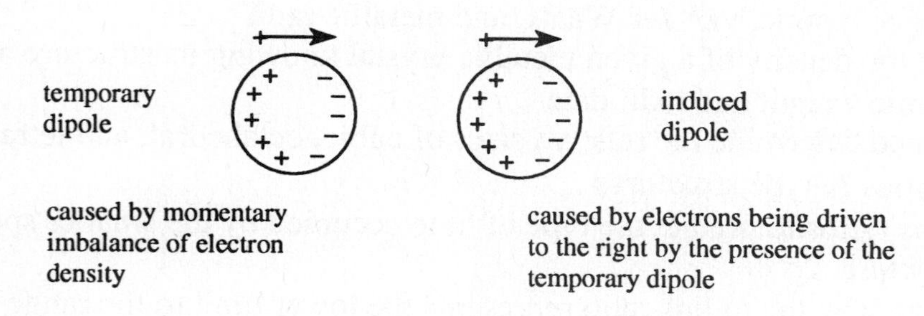

(b) Dipole-dipole forces are those among molecules possessing permanent dipole moments. One example might be among Cl-F molecules. The dotted lines represent dipole-dipole forces.

$^{\delta+}$Cl-F$^{\delta-}$ ··· $^{\delta+}$Cl-F$^{\delta-}$ ··· $^{\delta+}$Cl-F$^{\delta-}$ ··· $^{\delta+}$Cl-F$^{\delta-}$ ··· $^{\delta+}$Cl-F$^{\delta-}$···· $^{\delta+}$Cl-F$^{\delta-}$····

(c) Hydrogen bonds are special cases of dipole-dipole forces involving molecules in which a hydrogen atom is covalently bonded to an electronegative atom such as fluorine, oxygen, nitrogen, or even chlorine. One example might be among molecules of hydrogen fluoride. Here the dotted lines represent hydrogen bonds.

$^{\delta+}$H-F$^{\delta-}$···· $^{\delta+}$H-F$^{\delta-}$··· $^{\delta+}$H-F$^{\delta-}$ ··· $^{\delta+}$H-F$^{\delta-}$···· $^{\delta+}$H-F$^{\delta-}$···· $^{\delta+}$H-F$^{\delta-}$···

7.5. (a) does not account for all the space in the lattice; (b) is good but does not reflect the macroscopic structure of the crystal; (c) is best because it does reflect the macroscopic structure; (d) also does not reflect the macroscopic structure.

7.7. A 8 x 1/8 = 1
B 6 x 1/2 = 3

Therefore, AB_3 is the empirical formula. Face-centered cubic is not the most accurate designation because fcc implies that all spheres are identical to each other.

7.9. The twelve atoms which form the hexagonal prismatic shape of the unit cell are 1/6 in the unit cell yielding two atoms. The two atoms in the center of the hexagonal faces are 1/2 in the cell yielding an additional atom. Three atoms are essentially fully within the unit cell. The total is six atoms in the cell.

7.11. (a) To determine the volume of the hcp unit cell:

The base of the hexagonal prism is shown at right, where each edge is equal to 2r. By drawing the three diagonals of the hexagon (dashed lines), we can imagine the base to consist of <u>six</u> equilateral triangles of edge 2r (see separate triangle). The area of such a base is obtained by first obtaining the height of a component triangle ($r\sqrt{3}$, see below right) and then the area of the triangle. Finally, the total area of the hexagon is six times the area of one triangle.

$$A = \left[6 \times \frac{(2r)(r\sqrt{3})}{2}\right] = 6r^2\sqrt{3}$$

$$(2r)^2 = r^2 + x^2$$
$$x^2 = 3r^2$$
$$x = r\sqrt{3}$$

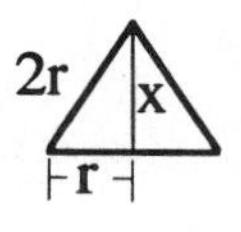

The height of the prism is just twice the height of a regular tetrahedron formed by the spheres of radius r. The height of a tetrahedron of edge length l is $l\sqrt{6}/3$ (see diagram at right).

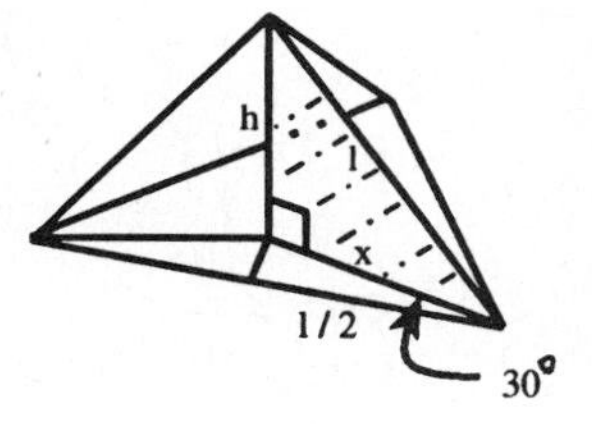

$$\cos 30 = \frac{1/2}{x} = \frac{\sqrt{3}}{2}, \quad \text{therefore, } x = 1/\sqrt{3}$$

Applying the Pythagorean theorem on the shaded triangle,

$$h^2 + x^2 = l^2, \quad \text{therefore, } h^2 + (l/\sqrt{3})^2 = l^2 \text{ and } h = (\sqrt{6}/3)l$$

Since $l = 2r$, the height of the tetrahedron is $(2r\sqrt{6})/3$.

It follows that the volume of the hexagonal prismatic unit cell is equal to the height of the prism $\left(\frac{4r\sqrt{6}}{3}\right)$ times the area of the base $(6r^2\sqrt{3})$.

$$V = \left(\frac{4r\sqrt{6}}{3}\right)(6r^2\sqrt{3}) = (24\sqrt{2})r^3$$

(b) The fraction of space occupied $= \dfrac{6\,(4\pi r^3/3)}{(24\sqrt{2})r^3} = \pi/(3\sqrt{2}) = 0.74$

7.13. Aluminum is face-centered cubic.

$$\text{Density} = 2.70 \text{ g/cm}^3 = \frac{(4 \text{ atoms})\left(\dfrac{26.98 \text{ g/mol}}{6.02 \times 10^{23} \text{ atoms/mol}}\right)}{l^3 \text{ cm}^3}$$

$$l = 4.05 \times 10^{-8} \text{ cm or } 4.05\text{Å}$$

7.15. Krypton is cubic close-packed or face-centered cubic.
(a) 4 atoms per unit cell
(b)

$$\text{Density} = 3.5\ \text{g/cm}^3 = \frac{(4\ \text{atoms})\left(\dfrac{83.80\ \text{g/mol}}{6.02 \times 10^{23}\ \text{atoms/mol}}\right)}{l^3\ \text{cm}^3}$$

$l = 5.4 \times 10^{-8}$ cm or 5.4Å

(c) $2d = 4r = l\sqrt{2}$, therefore $r = l\sqrt{2}/4 = 5.4\sqrt{2}/4 = 1.9$Å

7.17. (a) triangular hole

$$\cos 30° = \frac{l/2}{x} = \frac{\sqrt{3}}{2}$$

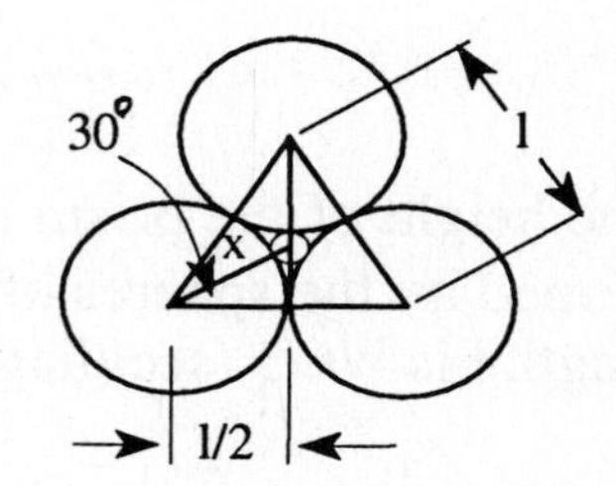

$x = l/\sqrt{3}$

Here, $x = r^+ + r^-$, $l = 2r^-$

$\therefore r^+ + r^- = 2r^-/\sqrt{3}$, therefore, $r^+/r^- = 0.15$

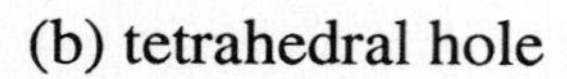
(b) tetrahedral hole

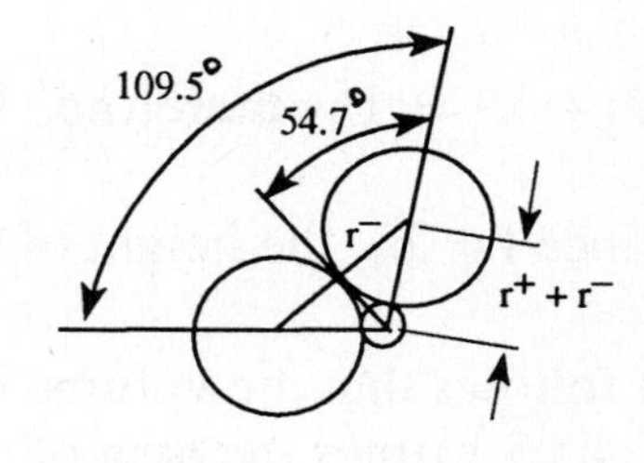

Consider two of the larger spheres forming the hole and the smaller sphere in the hole (as shown at right).

$$\sin 54.7° = \frac{r^-}{r^+ + r^-} = 0.816 \text{ which yields } r^+/r^- = 0.225$$

7.19.

(a) Zinc blende [see Figure 7.22(c)]
The X atoms are in a cubic close-packed unit cell, therefore there are 4 X atoms per unit cell [= 8(1/8) + 12(1/4)]. The M atoms are in half the tetrahedral holes, therefore there are 4 M atoms per unit cell. It follows that there are 4 MX formula units per unit cell.

(b) Wurtzite [see Figure 7.22(d)]
The X atoms are in a hexagonal close-packed unit cell, therefore there are 6 X atoms per unit cell (see solution to Problem 7.9). The M atoms are in half the tetrahedral holes, therefore there are 6 M atoms per unit cell. It follows that there are 6 MX formula units per unit cell.

7.21. In fluorite, the body diagonal equals $4r(Ca^{2+}) + 4r(F^-)$ even though in a given unit cell the center position is, in fact, empty. The body diagonal also equals $l\sqrt{3}$ (see Figure 7.9) where l is the length of the edge of the unit cell.

Therefore, $l\sqrt{3} = 4r(Ca^{2+}) + 4r(F^-) = 4(d_{Ca\text{-}F}) = 4(2.37\,Å)$

$$l = \frac{4(2.37)}{\sqrt{3}} = 5.47\,Å$$

$$\text{Density} = 3.18\ g/cm^3 = \frac{(4\ \text{atoms})\left(\dfrac{40.08\ g/mol}{N\ atoms/mol}\right) + (8\ \text{atoms})\left(\dfrac{19.00\ g/mol}{N\ atoms/mol}\right)}{(5.47 \times 10^{-8})^3}$$

Solving for N yields 6.00×10^{23} atoms/mol.

7.23. All radii in Ångstrom units for coordination number = 6

(a) $\frac{r(Sr^{2+})}{r(Cl^-)} = \frac{1.32}{1.67} = 0.79$ predicts cn = 8

correct, $SrCl_2$ assumes fluorite structure

(b) $\frac{r(Li^+)}{r(O^{2-})} = \frac{0.90}{1.26} = 0.71$ predicts cn = 6

incorrect, Li_2O assumes antifluorite structure

(c) $\frac{r(O^{2-})}{r(K^+)} = \frac{1.26}{1.52} = 0.83$ predicts cn = 8

correct, K_2O assumes antifluorite structure

(d) $\frac{r(Sn^{4+})}{r(S^{2-})} = \frac{0.83}{1.70} = 0.49$ predicts cn = 6

correct, SnS_2 assumes CdI_2 structure

(e) $\frac{r(Mg^{2+})}{r(F^-)} = \frac{0.86}{1.19} = 0.72$ predicts cn = 6

correct, MgF_2 assumes rutile structure

(f) $\frac{r(Mg^{2+})}{r(I^-)} = \frac{0.86}{2.06} = 0.42$ predicts cn = 6

correct, MgI_2 assumes CdI_2 structure

7.25. NaCl structure

Cl^- ions are in a face-centered cubic or cubic close-packed (abcabc) structure.

Na^+ ions are in all the octahedral holes.

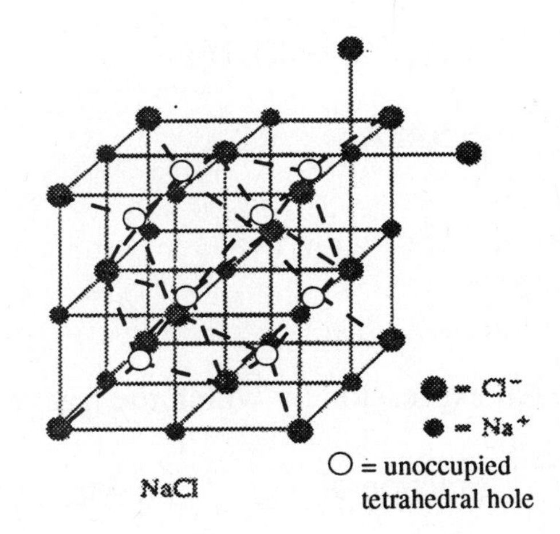

7.27. The occupied tetrahedral holes also form a tetrahedron in the wurtzite structure.

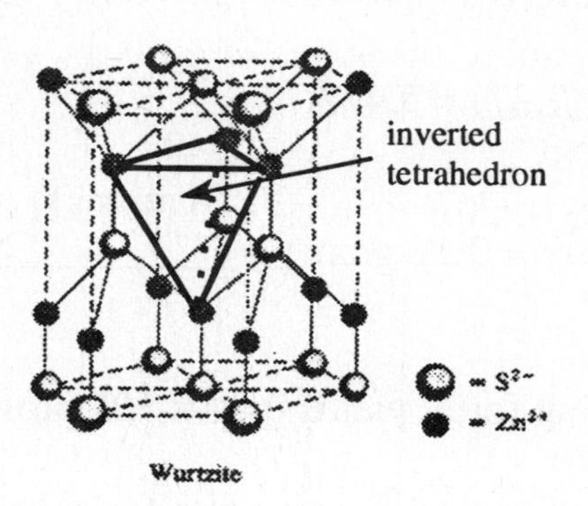

7.29. Since the arsenic atoms form a hexagonal close-packed structure, there are 6 As atoms per unit cell. The nickel atoms would occupy all 6 of the octahedral holes and therefore there would be 6 NiAs formula units per cell.

7.31. (a)

$$\text{Density} = 2.468 \text{ g/cm}^3 = \frac{(4 \text{ atoms})\left(\frac{39.10 \text{ g/mol}}{6.022 \times 10^{23} \text{ atoms/mol}}\right) + (4 \text{ atoms})\left(\frac{19.00 \text{ g/mol}}{6.022 \times 10^{23} \text{ atoms/mol}}\right)}{l^3 \text{ cm}^3}$$

Solving for l yields 5.387×10^{-8} cm (or 5.387Å).

We could also calculate a value for l using the figure given at the right.

Here $l = 2r(K^+) + 2r(F^-)$

$= 2(1.52) + 2(1.19)$

$= 5.42$Å

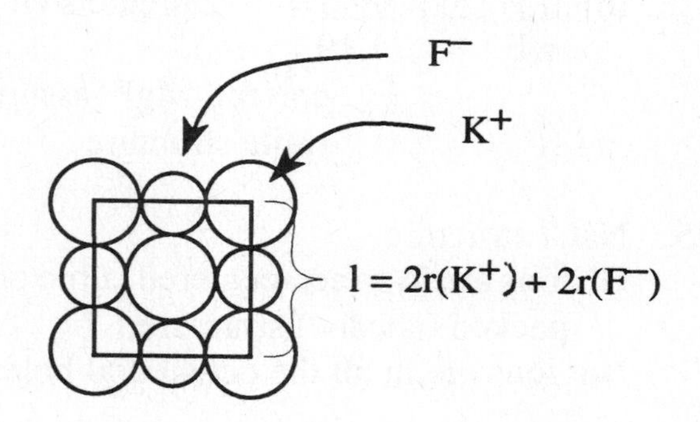

Note: K^+ and F^- are in contact but we cannot assume that the K^+'s are

(b) An expression by which the fraction of space occupied could be calculated is given below.

$$\text{Fraction of space occupied} = \frac{4\left(\frac{4\pi r_{K^+}^3}{3}\right) + 4\left(\frac{4\pi r_{F^-}^3}{3}\right)}{l^3}$$

(c) At first one might suspect that the fraction would be greater than 0.74, because the fluoride ions have been added to the fcc unit cell of potassium cations. On the other hand, the K^+'s are now separated. An actual calculation shows that this fraction comes out **less than 0.74**.

7.33. (a) W atoms at cube corners: 8 x 1/8 = 1 atom/unit cell
O atoms at centers of the cube edges: 12 x 1/4 = 3 atoms/unit cell
Na atoms at cube centers: 1 x 1 = 1 atom/unit cell

Therefore, the empirical formula is $NaWO_3$.

(b)

$$\text{Density} = \frac{(1\text{ atom})\left(\frac{23.0\text{ g/mol}}{N_A\text{ atoms/mol}}\right) + (1\text{ atom})\left(\frac{183.9\text{ g/mol}}{N_A\text{ atoms/mol}}\right) + (3\text{ atoms})\left(\frac{16.0\text{ g/mol}}{N_A\text{ atoms/mol}}\right)}{(3.86 \times 10^{-8})^3\text{ cm}^3}$$

where $N_A = 6.02 \times 10^{23}$

$$\text{Density} = \frac{4.23 \times 10^{-22}\text{ grams}}{5.75 \times 10^{-23}\text{ cm}^3} = 7.36\text{ g/cm}^3$$

7.35. MgO, according to Table 7.8, has the rock salt structure. Since the cations and anions can be assumed to be contact along a cell edge (but note that the face-centered cubic anions cannot be assumed to be contact along the face diagonal -- see the figure in the solution to Problem 7.31), it follows that l, the cell edge, is just equal to $2r(Mg^{2+}) + 2r(O^{2-})$.

$$l = 2(0.86) + 2(1.26) = 4.24\text{Å}$$

Volume of unit cell = $l^3 = (4.24\text{Å})^3 = 76.2\text{Å}^3$ (or 7.62×10^{-23} cm^3)

$$\text{Density} = \frac{\frac{4(24.3) + 4(16.0)\text{ grams}}{6.02 \times 10^{23}}}{7.62 \times 10^{-23}\text{ cm}^3} = 3.51\text{ g/cm}^3$$

The actual density is listed as 3.58 g/cm^3.

7.37. A model of the nickel arsenide structure would show that there is very little shielding between layers of nickel atoms. If these sites were to be occupied by ions, a high degree of repulsion would significantly destabilize the structure. Compounds with a high degree of ionic character would be more stable in the NaCl structure.

7.39.

$\frac{r(NH_4^+)}{r(I^-)} = \frac{1.37}{2.06} = 0.665$	$\frac{r(NH_4^+)}{r(Cl^-)} = \frac{1.37}{1.67} = 0.820$	$\frac{r(F^-)}{r(NH_4^+)} = \frac{1.19}{1.37} = 0.869$
indicates cn = 6; correct, NH_4I has a rock salt structure	indicates cn = 8; correct, NH_4Cl has a CsCl structure	indicates cn = 8; incorrect, NH_4F has a wurtzite structure

7.41. Assuming the Na^+ and SbF_6^- ions to be in contact along the cell edge, we can set up the following expression for the density of $NaSbF_6$. (FW of SbF_6^- = 235.75 g/mole).

$$\text{Density} = \frac{\dfrac{4(22.99) + 4(235.75)\text{ grams}}{6.02 \times 10^{23}}}{[2(1.16 \times 10^{-8}) + 2r(SbF_6^-)]^3\text{ cm}^3} = 4.37\text{ g/cm}^3$$

Solving for $r(SbF_6^-)$ yields a value of 2.50×10^{-8} cm or 2.50Å

7.43. Transition metal oxides are more frequently nonstoichiometric because these elements usually have a variety of oxidation states, more so than nontransition metal oxides. With more oxidation states, the charge of missing ions can be made up by oxidizing some of the remaining ions to a higher stable oxidation state.

7.45. The oxide ions are face-centered cubic, therefore there are 8(1/8) + 6(1/2) = 4 per unit cell. There are 8 tetrahedral holes in an fcc unit cell (1 for each corner of the cube). If 1/8 of these are occupied by the A^{II} cations, there is 1 A^{II} per unit cell. There are 1 + 12(1/4) = 4 octahedral holes in an fcc unit cell. If one half of these are occupied by B^{III} cations, there are 2 B^{III}'s per unit cell. It follows that the stoichiometry of such a unit cell and therefore the entire compound is $A^{II}B_2^{III}O_4$.

7.47. Mn_3O_4 is made up of Mn(II) and Mn(III), that is, it is $Mn^{III}_2Mn^{II}O_4$. In a normal spinel structure, the Mn(III) ions would occupy half the octahedral holes in the face-centered cubic array of oxide ions while the Mn(II) ions would occupy one eighth of the tetrahedral holes. In an inverse spinel the Mn(II) would exchange places with one half of the Mn(III) cations. Manganese has an electronic configuration of $[Ar]4s^23d^5$, therefore Mn(II) is $[Ar]3d^5$ and Mn(III) is $[Ar]3d^4$. In a high-spin crystal field, the CFSE of a d^5 ion is zero for both tetrahedral and octahedral cases. (See Figure 7.27(b) for these calculations.) The calculations of the CFSE for a d^4 ion in both octahedral and tetrahedral fields is shown below. We see that the Mn(III) d^4 ion clearly prefers an octahedral environment. Therefore, since the A^{III} ions occupy the octahedral holes in a normal spinel, this is the structure that the compound is expected to assume.

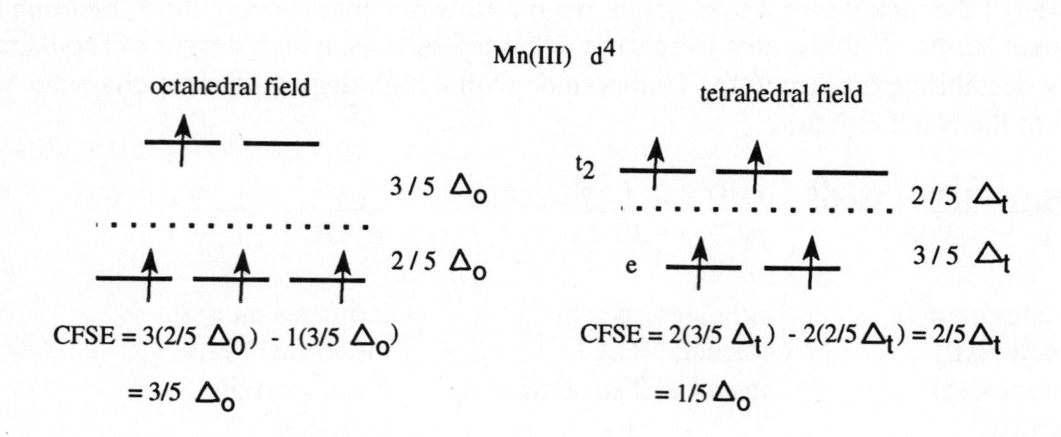

Chapter 8
Solid State Energetics

The sections and subsections in this chapter are listed below.

8.1 Lattice Energy: A Theoretical Evaluation
8.2 Lattice Energy: Thermodynamic Cycles
  Electron Affinities
  Heats of Formation for Unknown Compounds
  Thermochemical Radii
8.3 Lattices Energies and Ionic Radii: Connecting Crystal Field Effects with Solid State Energetics

Chapter Objectives

You should be able to

- define and discuss both the electrostatic and short-range repulsive forces that contribute to the overall lattice energy of an ionic compound
- derive and discuss the components of the Born-Landé equation for the lattice energy of an ionic compound
- discuss the role of charge density as a factor in determining the magnitude of the lattice energy of a compound
- discuss the role of Kapustinskii equation in estimating the lattice energy of an ionic compound
- write a Born-Haber thermodynamic cycle that can be used to derive a value of the lattice energy of an ionic compound
- discuss the conditions under which significant covalent contributions to lattice energy would be expected
- use thermodynamic cycles and the Born-Landé or Kapustinskii equations to estimate values of electron affinity, heats of formation for unknown compounds, and thermochemical radii
- account qualitatively and quantitatively for the role of crystal field effects in determining the cationic radii and lattice energies of transition metal salts

8.1. (a) $\frac{r(Rb^+)}{r(Br^-)} = \frac{1.66}{1.82} = 0.91$

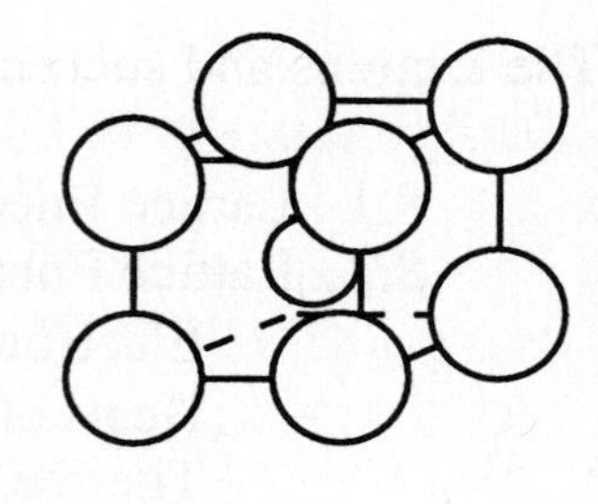

This radius ratio is consistent with a coordination number of eight. Therefore, a reasonable* unit cell would be that of cesium chloride, that is, a simple cubic array of bromide ions with rubidium cations in the center of each cell. A sketch of such a cell is found at the right.
(*The actual unit cell of RbBr is rock salt, but CsCl is a reasonable structure nevertheless for this compound.)

(b) $U_o = 1389 \frac{Z^+Z^-M}{r_o}\left(1 - \frac{1}{n}\right) = \frac{1389\,(+1)(-1)\,(1.763)}{(1.66 + 1.82)}\left(1 - \frac{1}{10}\right) = -\,633$ kJ/mol

8.3. There are two reasonable approaches to this problem. (1) Since we do not know the crystal structure, we can simply use the Kapustinskii equation and solve for $r(Fr^+)$.

$- 632\text{kJ/mol} = \frac{1202(2)(+1)(-1)}{r_o}\left(1 - \frac{0.345}{r_o}\right)$

Solving for r_o yields a value of 3.42Å; 3.42Å = $r(Fr^+) + r(Cl^-) = r(Fr^+) + 1.67$
Therefore, $r(Fr^+)$ = 1.75Å

(2) A second approach: since all the other alkali halides assume the rock salt structure, it would be logical to assume that FrCl does also. In that case we can use the Born-Landé equation with n = 14 for Fr^+, which has an [Rn] shell (the value of 14 was obtained by extrapolating from the data given in Table 8.2), n = 9 for Cl^- which has an [Ar] shell. The average value of n is 11.5.

$U_o = 1389\frac{(+1)(-1)\,(1.748)}{r_o}\left(1 - \frac{1}{11.5}\right) = -\,632$ kJ/mol; Solving for r_o yields a value of 3.51Å.

$r(Fr^+) + r(Cl^-) = r(Fr^+) + 1.67$ = 3.51Å; therefore, $r(Fr^+)$ = 1.84Å.

8.5. (a) $\frac{r(Bk^{4+})}{r(O^{2-})} = \frac{0.97}{1.26} = 0.77$ which is consistent with a coordination number of 8 and a fluorite lattice.

$U_o = 1389\frac{Z^+Z^-M}{r_o}\left(1 - \frac{1}{n}\right) = \frac{1389\,(+4)(-2)(2.519)}{(0.97 + 1.26)}\left(1 - \frac{1}{10.5}\right) = -\,11{,}400$ kJ/mol

The value of n is obtained by averaging n = 7 for O^{2-}, which has an [Ne] shell and a value of n = 14 for Bk^{4+}, which has an [Rn]$4f^7$ electronic configuration. n = 14 is obtained by extrapolation of the data in Table 8.2.

(b) For a CdI_2 structure, Z = 2.191 instead of 2.519 for fluorite.

Now the lattice energy comes out to be -9880kJ/mol, about a 13% difference.

(c) $U_{Kap} = \dfrac{1202(3)(+4)(-2)}{0.97 + 1.26}\left(1 - \dfrac{0.345}{2.23}\right) = -\ 10{,}900 \text{ kJ/mol}$

The Kapustinskii value lies in between the above two values. Without knowing for sure what the crystal structure of BkO_2 is, the U_{Kap} seems to be a reasonable approximation of the lattice energy.

8.7. $\dfrac{r(Mg^{2+})}{r(S^{2-})} = \dfrac{0.86}{1.70} = 0.51$ $\dfrac{r(Li^{+})}{r(Br^{-})} = \dfrac{0.90}{1.82} = 0.50$

Given the above radius ratios, it would seem reasonable that these two compounds might exhibit the same crystal structures. (This conclusion is also supported by the data in Table 7.8, which indicates that both assume the rock salt structure.) The primary difference between these two compounds is the charges, +1/-1 for LiBr and +2/-2 for MgS. These charges would result in the lattice energy of MgS being approximately four times that of LiBr making it more difficult to break apart. Held together by greater electrostatic forces, MgS should be harder and higher melting.

8.9.

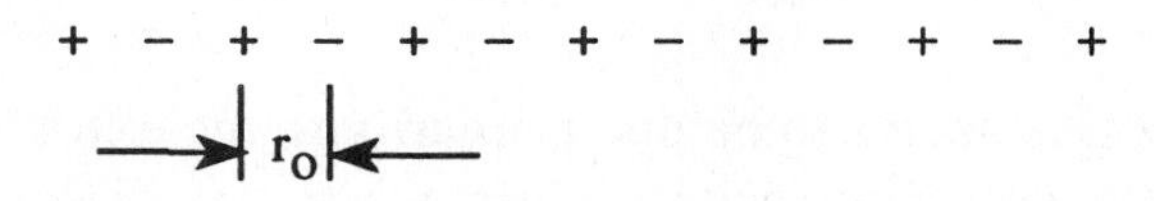

$$E = -\frac{2e^2}{r_o} + \frac{2e^2}{2\,r_o} - \frac{2e^2}{3\,r_o} + \frac{2e^2}{4\,r_o} + \ldots$$

$$E = -\frac{e^2}{r_o}[2(1 - 1/2 + 1/3 - 1/4 + \ldots)] = -\frac{e^2\,[2(\ln 2)]}{r_o}$$

(The expression in the parentheses is called the "alternating harmonic series." It converges to a value of ln 2.)

So the Madelung constant in this case, M = 2 ln 2 = 1.39

8.11.

Li(s) + 1/2 F_2(g) —ΔH^o_f→ LiF(s)

Li(s) —ΔH_{subl}→ Li(g)

1/2 F_2(g) —1/2 D→ F(g)

F(g) —EA→ F^-(g) + Li^+(g)

Li(g) —IE→ F^-(g) + Li^+(g)

F^-(g) + Li^+(g) —U→ LiF(s)

$\Delta H^o_f \;= \Delta H_{subl} + IE + 1/2D + EA + U$
$U \;= \Delta H^o_f - \Delta H_{subl} - IE - 1/2D - EA$
$= -616.0 - 159.4 - 520.3 - 79.0 - (-328.0) = -\ 1046.7$ kJ/mol

Table 8.3 gives U(B-H) = - 1046.7kJ/mol

8.13.

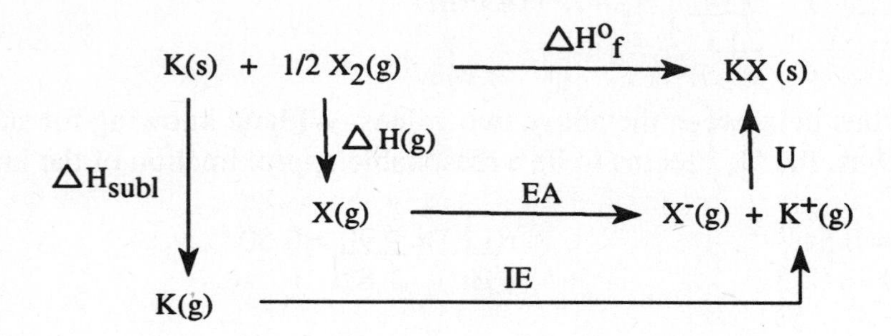

	U	=	ΔH°_f	$-\Delta H_{subl}$	-IE	$-\Delta H(g)$	-EA	All values (in kJ/mol) from Table 8.3
X = F	-784			const	const	-79.0	- (-328.0)	
X = Cl	-677			const	const	-121.7	- (-349.0)	
$\Delta_{F\text{-}Cl}$	-107			const	const	+42.7	-21.0	

The higher reactivity of the fluorine seems to be due primarily to the extra lattice energy of the fluorides (represented here by the potassium salts). More energy (107kJ/mol) is released when a representative fluoride-containing lattice is formed than when a chloride lattice is. The lower value of ΔH(g) for F(g) would also seem to contribute to fluorine's greater reactivity. (It does not take as much energy to break the F-F bond as it does to break the Cl-Cl bond.) The difference in electron affinities is a factor that makes the chloride salts more stable than the fluorides (that is, makes chlorine more reactive), but this is a comparatively small factor.

8.15.

$1/2\, Cl_2(g) + 1/2\, F_2(g) \xrightarrow{\Delta H^\circ_f} ClF(g)$

$1/2\, D_{Cl\text{-}Cl}$ ↓ ↓ $1/2\, D_{F\text{-}F}$

$Cl(g) + F(g)$ → ClF(g): $-D_{Cl\text{-}F}$

$$\Delta H^\circ_f = 1/2 D_{Cl\text{-}Cl} + 1/2 D_{F\text{-}F} - D_{F\text{-}Cl}$$

$$1/2 D_{F\text{-}F} = \Delta H^\circ_f - 1/2 D_{Cl\text{-}Cl} + D_{Cl\text{-}F}$$

$$= -56.1 - 1/2(243.4) + 246.4$$

$$= 68.6 \text{kJ/mol}$$

$$D_{F\text{-}F} = 137.2 \text{kJ/mol}$$

8.17. We want to calculate a value for the $\Delta H°$ for the following equation:

CsCl (ordinary form) $\rightarrow$ CsCl (rock salt form)

When we calculate the $\Delta H°_f$ for these two compounds, all the factors will be the same except the lattice energies. In fact, the ΔH of the above process can be calculated as follows:

$$\Delta H = \Delta H°_f(\text{rock salt form}) - \Delta H°_f(\text{ordinary form})$$

$$= U(\text{rock salt form}) - U(\text{ordinary form})$$

The lattice energies will differ only because the Madelung constants differ. Knowing this, we can calculate the Born-Landé lattice energies for each form and proceed to calculate ΔH.

$$U_o(\text{rock salt form}) = 1389\frac{Z^+Z^-M}{r_o}\left(1 - \frac{1}{n}\right) = 1389\frac{(+1)(-1)(1.748)}{(1.81 + 1.67)}\left(1 - \frac{1}{10.5}\right) = -631\text{kJ/mol}$$

$$U_o(\text{ordinary form}) = 1389\frac{Z^+Z^-M}{r_o}\left(1 - \frac{1}{n}\right) = 1389\frac{(+1)(-1)(1.763)}{(1.81 + 1.67)}\left(1 - \frac{1}{10.5}\right) = -637\text{kJ/mol}$$

$$\Delta H = -631\text{kJ/mol} - (-637\text{kJ/mol}) = 6\text{kJ/mol}$$

8.19. (a)

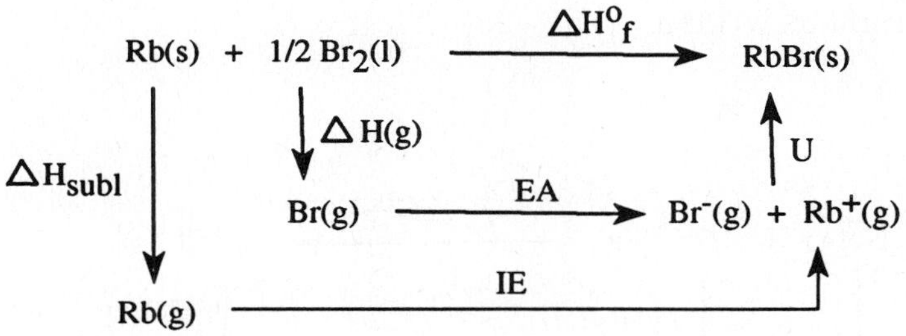

$Rb(s)$ + 1/2 $Br_2(l)$ —ΔH^o_f→ $RbBr(s)$

ΔH_{subl}

$\Delta H(g)$

U

$Br(g)$ —EA→ $Br^-(g)$ + $Rb^+(g)$

IE

$Rb(g)$

(b) From part (b) of Problem 8.1, we know that $U_o^{B\text{-}L} = -633\text{kJ/mol}$

$$EA = \Delta H°_f - \Delta H_{sub} - IE - \Delta H(g) - U$$

$$EA = -394.6 - 80.9 - 403.1 - 111.9 - (-633) = -358\text{kJ/mol}$$

This result is somewhat different from the value given in Table 8.3 (-324.7) because RbBr actually assumes the rock salt (NaCl) structure not the CsCl structure used in Problem 8.1.

8.21. (a) $U_o = 1389\frac{Z^+Z^-M}{r_o}\left(1 - \frac{1}{n}\right) = 1389\frac{(+2)(-2)(1.641)}{0.74 + 1.70}\left(1 - \frac{1}{9}\right) = -3320\text{kJ/mol}$

[The value of $r(Zn^{2+})$ is for coordination number 4.]

(b)

$Zn(s) + S(s) \xrightarrow{\Delta H^o_f} ZnS(s)$

ΔH_{subl}: $Zn(s) \rightarrow Zn(g)$; $\Delta H(g)$: $S(s) \rightarrow S(g)$; $S(g) \xrightarrow{EA^1} \xrightarrow{EA^2} S^{2-}(g) + Zn^{2+}(g)$; $Zn(g) \xrightarrow{IE^1 + IE^2}$; U: $\rightarrow ZnS(s)$

$\Delta H^\circ_f \;=\; \Delta H_{subl} + IE^1 + IE^2 + \Delta H(g) + EA^1 + EA^2 + U$

$EA^2 \;=\; \Delta H^\circ_f - \Delta H_{subl} - IE^1 - IE^2 - \Delta H(g) - EA^1 - U$

$EA^2 \;=\; -192.6 - 130.8 - 906.4 - 1733 - 278.8 - (-200.4) - (-3320) = 280\text{kJ/mol}$

(Note: there are only two significant figures in this value.)

8.23. $2CaCl(s) \rightarrow CaCl_2(s) + Ca(s)$

$\Delta H = \Delta H^\circ_f [CaCl_2(s)] + \Delta H^\circ_f [Ca(s)] - 2\Delta H^\circ_f [CaCl(s)]$

$\Delta H = -792 + 0 - 2(-130) = -532\text{kJ/mol}$

Yes, we would predict that this reaction would probably occur. It has a negative heat of reaction and since there are two moles of solids on either side of the equation, the ΔS is probably not appreciably negative. These factors would make ΔG negative and the reaction (under standard state conditions) spontaneous from left to right as written.

8.25.

$Ne(g) + 1/2\, Cl_2(g) \xrightarrow{\Delta H^o_f} Ne^+Cl^-(s)$

$\Delta H(g)$: $1/2\, Cl_2(g) \rightarrow Cl(g)$; $Cl(g) \xrightarrow{EA} Cl^-(g) + Ne^+(g)$; IE: $Ne(g) \rightarrow Ne^+(g)$; U: $\rightarrow Ne^+Cl^-(s)$

$\Delta H^\circ_f \;=\; IE + \Delta H(g) + EA + U$

$$U_{Kap} = 1202 \frac{\nu Z^+ Z^-}{r_o}\left(1 - \frac{0.345}{r_o}\right) = 1202 \frac{(2)(+1)(-1)}{3.07}\left(1 - \frac{0.345}{3.07}\right) = -695\text{kJ/mol}$$

Table 7.2 gives a value of 1.54Å for the van der Waals radius of neon. Removing one electron from a filled shell should not decrease the size much at all. Let's arbitrarily use a value of 1.40Å for the radius of Ne^+. This combined with the ionic radius of Cl^- of 1.67Å gives a value for r_o of 3.07Å.

$\Delta H^\circ_f \;= 2080 + 121.7 + (-349.0) + (-695) = 1158\text{kJ/mol}$

NeCl would not form because the ionization energy of neon is too large. Not enough energy is released during the formation of the NeCl(s) lattice to make up for the energy needed to ionize an electron from the stable noble gas electronic configuration of neon.

8.27. The equation representing the reaction in which $CaCl_3$ might be produced from its constituent elements in their standard states would be as follows:

$$Ca(s) + 3/2\ Cl_2(g) \rightarrow CaCl_3(s)$$

The entropy change accompanying the above reaction would be negative since one and a half moles of gaseous reactants is going to zero moles of gaseous products. This negative $\Delta S_f°$, combined with the +1600kJ/mol calculated for the $\Delta H°_f$ of this compound, would always yield a positive value for $\Delta G°_f$. Given these values of $\Delta H°_f$ and $\Delta S_f°$, the entropy (which will always be negative) could never be used to force the formation of $CaCl_3$.

8.29. (a)

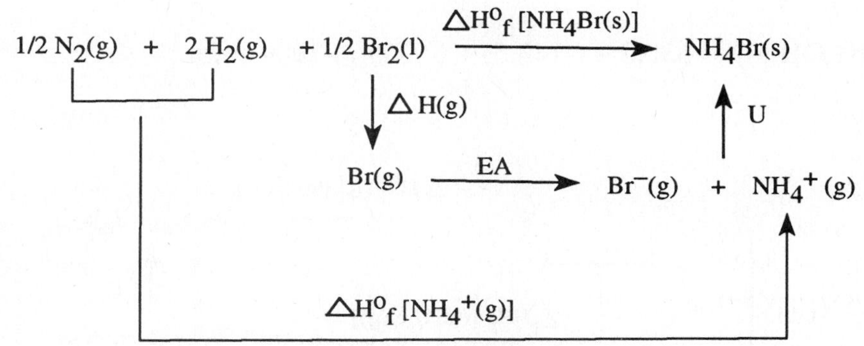

$$1/2\ N_2(g) + 2\ H_2(g) + 1/2\ Br_2(l) \xrightarrow{\Delta H^o_f [NH_4Br(s)]} NH_4Br(s)$$

$1/2\ Br_2(l) \xrightarrow{\Delta H(g)} Br(g)$

$Br(g) \xrightarrow{EA} Br^-(g)$

$1/2\ N_2(g) + 2\ H_2(g) \xrightarrow{\Delta H^o_f [NH_4^+(g)]} NH_4^+(g)$

$Br^-(g) + NH_4^+(g) \xrightarrow{U} NH_4Br(s)$

$$\Delta H°_f [NH_4Br(s)] = \Delta H°_f [NH_4^+(g)] + \Delta H(g) + EA + U$$
$$U = \Delta H°_f [NH_4Br(s)] - \Delta H°_f [NH_4^+(g)] - \Delta H(g) - EA$$
$$U = -270.3 - 630.2 - 111.9 - (-324.7) = -687.7 kJ/mol$$

(b) $$U_{Kap} = 1202 \frac{v Z^+ Z^-}{r_o}\left(1 - \frac{0.345}{r_o}\right) = 1202 \frac{(2)(+1)(-1)}{r_o}\left(1 - \frac{0.345}{r_o}\right) = -687.7 kJ/mol$$

Solving the above by the quadratic formula yields values of r_o of 3.11 and 0.39Å, of which only 3.11 is physically possible.

Since $r_o = r(NH_4^+) + r(Br^-) = 3.11$, therefore, $r(NH_4^+) = 3.11 - 1.82 = 1.29$Å.

This result is within 6% of the value given in Problem 7.39. This latter value is obtained by averaging the results of a number of thermochemical calculations involving the ammonium cation.

8.31. The standard heat of formation corresponds to the enthalpy change for the following reaction.

$$1/2 N_2(g) + 5/2\ H_2(g) + 1/2\ O_2(g) \rightarrow NH_4OH(s)$$

To calculate a value of the $\Delta H°_f[NH_4OH(s)]$, we must write a Born-Haber cycle as follows:

$$\frac{1}{2} N_2(g) + 2 H_2(g) + \frac{1}{2} O_2(g) + \frac{1}{2} H_2(g) \xrightarrow{\Delta H^o_f [NH_4OH(s)]} NH_4OH(s)$$

$\Delta H^o_f [NH_4^+(g)]$ ↓ $\Delta H^o_f [OH^-(g)]$ ↓ U ↑

$$NH_4^+(g) + OH^-(g)$$

$$\Delta H^\circ_f [NH_4OH(s)] = \Delta H^\circ_f [NH_4^+(g)] + \Delta H^\circ_f [OH^-(g)] + U$$

The lattice energy can be estimated using the Kapustinskii equation.

$$U_{Kap} = 1202 \frac{\nu Z^+ Z^-}{r_o}\left(1 - \frac{0.345}{r_o}\right) = 1202 \frac{(2)(+1)(-1)}{1.37 + 1.33}\left(1 - \frac{0.345}{1.37 + 1.33}\right) = -777 \text{kJ/mol}$$

Therefore, $\Delta H^\circ_f [NH_4OH(s)] = 630.2 + (-143.5) + (-777) = -290$kJ/mol

8.33.

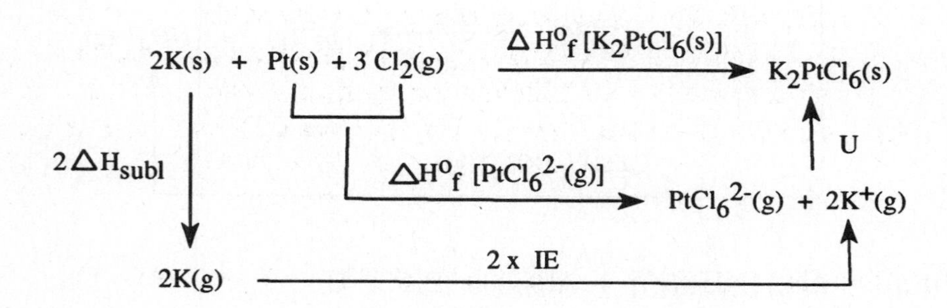

$$\Delta H^\circ_f [K_2PtCl_6(s)] = \Delta H^\circ_f [PtCl_6^{2-}(g)] + 2(\Delta H_{subl}) + 2(IE) + U$$
$$\Delta H^\circ_f [K_2PtCl_6(s)] = -774 + 2(89.2) + 2(418.9) + (-1468) = -1226 \text{kJ/mol}$$

8.35. $U_{Kap} = 1202 \frac{\nu Z^+ Z^-}{r_o}\left(1 - \frac{0.345}{r_o}\right) = 1202 \frac{(2)(+2)(-2)}{r_o}\left(1 - \frac{0.345}{r_o}\right) = -2911$kJ/mol

Solving for r_o yields values of 2.91 and 0.39Å, the latter value being physically impossible.
If $r_o = r(Ca^{2+}) + r(C_2^{2-}) = 1.14 + r(C_2^{2-}) = 2.91$Å, then $r(C_2^{2-}) = 1.77$Å.

8.37.

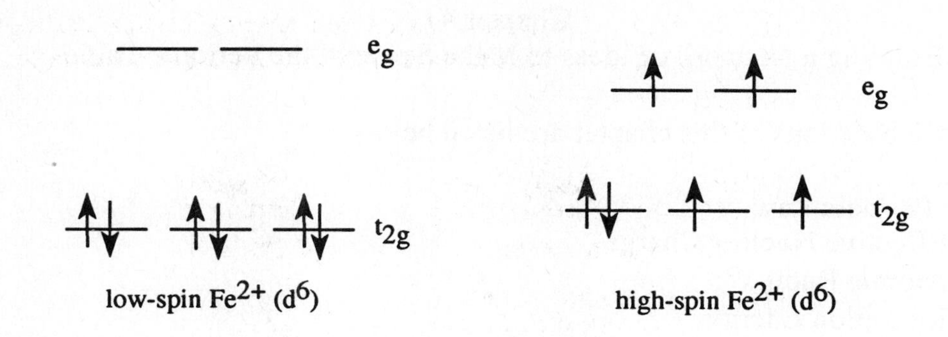

The high-spin Fe^{2+} has two electrons occupying the e_g orbitals that point directly at the octahedrally-placed ligands, therefore pushing these ligands back. The result is that the high-spin iron(II) is effectively larger than the low-spin.

8.39.

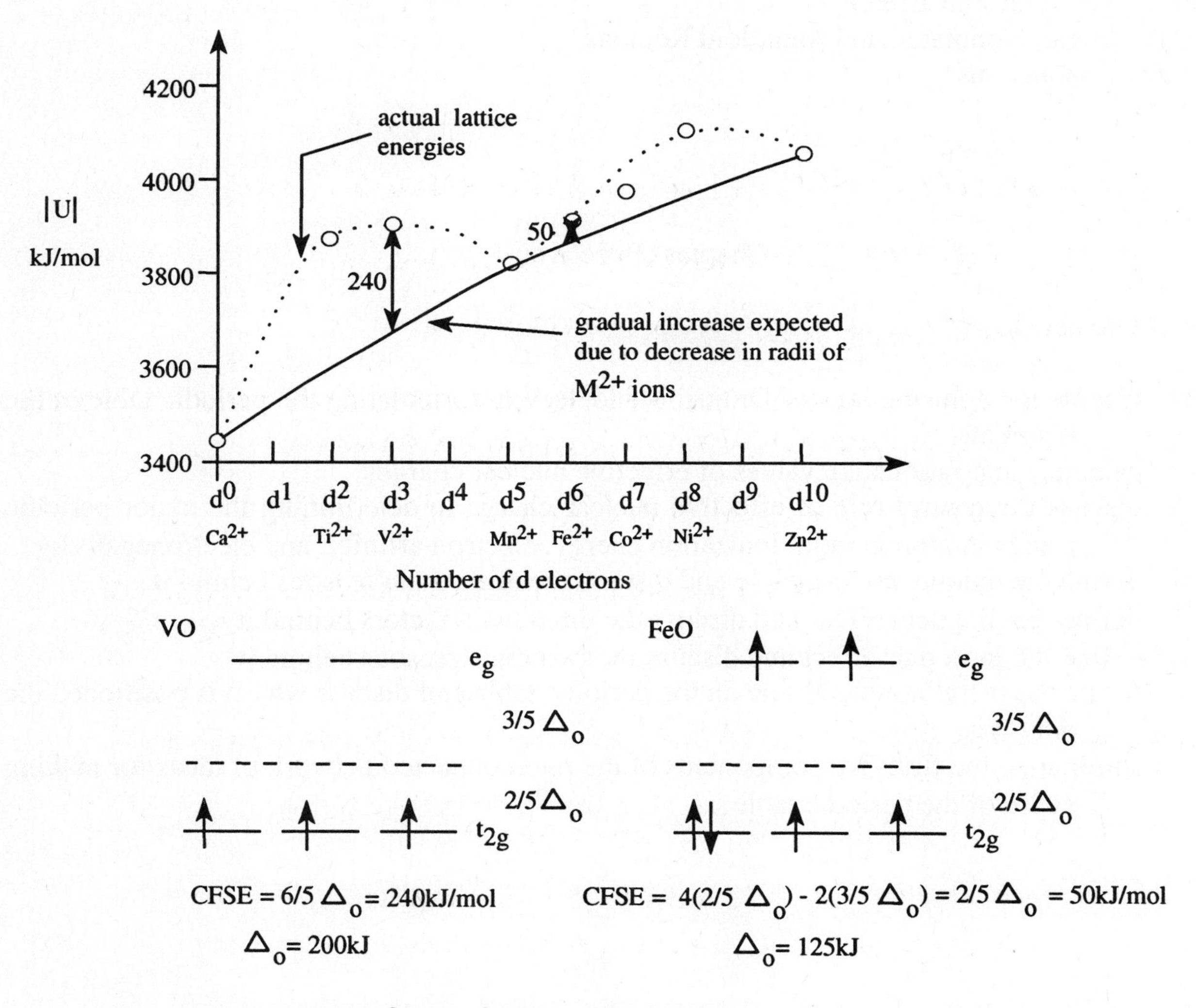

For MnO, the Mn^{2+} is $t_{2g}{}^3 e_g{}^2$, therefore CFSE = 0.

Chapter 9
Building a Network of Ideas to Make Sense of the Periodic Table

The sections and subsections of this chapter are listed below.

9.1 The Periodic Law
- Effective Nuclear Charge
- Atomic Radii
- Ionization Energy
- Electron Affinity
- Electronegativity

9.2 The Uniqueness Principle
- The Small Size of the First Elements
- The Increased Liklihood of π-bonding in the First Elements
- The Lack of Availability of d Orbitals in the First Elements

9.3 The Diagonal Effect

9.4 The Inert Pair Effect

9.5 Metal, Nonmetal, and Metalloid Regions

9.6 Conclusions

Chapter Objectives

You should be able to

- □ discuss the contributions of Dmitrii Mendeleev in formulating the periodic table of the elements
- □ calculate and rationalize values of effective nuclear charges
- □ discuss the central role of effective nuclear charge in determining the major periodic trends in atomic radii, ionization energy, electron affinity, and electronegativity
- □ define the uniqueness principle and discuss the three main reasons behind it
- □ define the diagonal effect and discuss the three main factors behind it
- □ define the inert pair effect and discuss the two main reasons behind it
- □ locate the metal/nonmetal line on the periodic table and discuss why it is positioned the way it is
- □ summarize the first five components of the interconnected network of ideas for making sense of the periodic table

9.1. The properties of eka-silicon (germanium) can be interpolated from those of silicon and tin (above and below it) and gallium and arsenic (to its left and right). Some reasonably predicted and actual values are given below.

	Actual for Ge	Predicted Vertically	Predicted Horizontally
atomic weight (u)	72.59	73.4	72.3
density (g/cm³)	5.32	4.8	5.8
melting point (°C)	937	823	421
boiling point (°C)	2830	2312	---
electronegativity	1.8	1.8	1.8
ionization energy (kJ/mol)	760	748	763
electron affinity (kJ/mol)	-118	-120	-56
atomic radius (Å)	1.39	---	---

9.3. Transition metals are often defined as elements that contain an incomplete d subshell. Zinc, cadmium, and mercury are d^{10} metals and therefore would not officially be classified under this definition.

9.5. Certainly nitrogen is a "choker or strangler" because we would assuredly be strangled if we tried breathing only nitrogen. Arsenic is a well-known poison but not a strangler. Phosphorus reacts spontaneously with oxygen in the air to form an oxide, but it is not a choker or strangler in and of itself.

9.7. According to Webster's New Universal Unabridged Dictionary, 2nd Edition, Dorset & Baber, 1979, the word halogen is from the Greek prefix *halos* for salt and *-gen* means to produce. Halogen, then, literally means "salt producer." Chlorine and the other Group 7A elements do produce salts when reacted with the alkali metals.

9.9. The plot of atomic volumes is another demonstration of the periodic properties of the elements. For example the maximum atomic volume in each period occurs at the alkali metals lithium, sodium, potassium, etc. Also we see that atomic volume tends to go through a minimum as a period of the elements is crossed, for example, from lithium to fluorine and sodium to chlorine.

9.11. (a) phosphorus $[Ne]3s^2 3p^3$; copper $[Ar]4s^1 3d^{10}$; arsenic $[Ar]4s^2 3d^{10} 4p^3$; thallium $[Xe]6s^2 4f^{14} 5d^{10} 6p^1$.

9.13. There is, of course, no such thing as a pseudonoble gas so in that sense the term "pseudonoble gas configuration" is somewhat misleading. The special stability of a noble gas electronic configuration is recognized to be a useful concept for rationalizing the existence of both the noble gases and various isoelectronic species, such as the Cl^- and Sr^{2+} ions, to cite just two examples. (The special stability of these electronic configurations is due to their resistance both to electron removal -- ionization -- and electron addition.) There are, however, many stable ionic species that do not have a noble gas electron configuration, but rather that plus a filled d or f subshell. (For example, the Br^-

ion or the mercuric ion, Hg^{2+}.) The stability of these species can be rationalized by speaking about the achievement of a pseudonoble gas configuration that is also resistant to both electron removal and addition.

9.15. (a) For an electron being added to the 3s orbital of Ne, $Z_{eff} = 10 - 10 = 0$. Here, all ten electrons of $1s^22s^22p^6$ are assumed to shield the 3s electron from the +10 nuclear charge. (b) For an electron being ionized from the 2p orbital of Ne, $Z_{eff} = 10 - 2 = 8$. Now only the $1s^2$ electrons shield the 2s or 2p electrons from the +10 nucleus. These results show that neon is resistant to the addition of an electron ($Z_{eff} = 0$) and to the removal of an electron ($Z_{eff} = +8$). Accordingly, the neon electron configuration is particularly stable.

9.17.

Element	Electron Configuration	Z	σ (a)	Z_{eff}(a)	σ(b)	Z_{eff} (b)
Li	$[He]1s^1$	3	2	1	2(0.85) = 1.70	1.30
Be	$[He]2s^2$	4	2	2	2(0.85) + 1(0.35) = 2.05	1.95
B	$[He]2s^22p^1$	5	2	3	2(0.85) + 2(0.35) = 2.40	2.60
C	$[He]2s^22p^2$	6	2	4	2(0.85) + 3(0.35) = 2.75	3.25
N	$[He]2s^22p^3$	7	2	5	2(0.85) + 4(0.35) = 3.10	3.90
O	$[He]2s^22p^4$	8	2	6	2(0.85) + 5(0.35) = 3.45	4.55
F	$[He]2s^22p^5$	9	2	7	2(0.85) + 6(0.35) = 3.80	5.20
Ne	$[He]2s^22p^6$	10	2	8	2(0.85) + 7(0.35) = 4.15	5.85

As shown below, plots of both the σ (a) (solid line) and σ(b) (dashed line) against atomic number demonstrate that the effective nuclear charge increases linearly across the period.

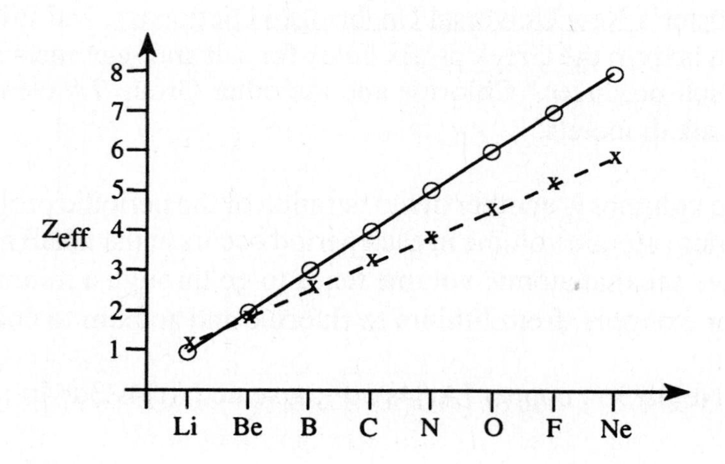

9.19. For the purposes of calculating the effective nuclear charge using Slater's rules, the electronic configuration of copper is written as follows: $1s^22s^22p^63s^23p^63d^{10}4s^1$.

For a 4s electron of copper, $Z_{eff} = 29 - 10(0.85) - 18(1.00) = 2.50$
For a 3d electron of copper, $Z_{eff} = 29 - 9(0.35) - 18(1.00) = 7.85$

The effective nuclear charge holding a 4s electron in copper is much less than that holding a 3d electron. Therefore, we have a rationale for why the 4s electrons are ionized before the 3d.

9.21.

Lithium	$Z_{eff} = 3 - 2 = 1$
Sodium	$Z_{eff} = 11 - 10 = 1$
Potassium	$Z_{eff} = 19 - 18 = 1$

Ionization energy depends on both Z_{eff} and the distance from the valence electron to that effective nuclear charge. Going down the group, the valence ns electrons are farther and farther away from the same (+1) Z_{eff} and therefore are easier to remove.

9.23. The energy of an ns electron is always less than that of the corresponding np electron. This difference is because s orbitals penetrate through the inner-core electrons to the nucleus better than the np orbitals do. The fact that Slater's rules treat the ns and np orbitals as members of the same group represents an oversimplification. To reflect the fact that ns orbitals are lower in energy, the contribution to the shielding constant should be a little less for groups of orbitals inside an ns orbital than the contribution for groups inside an np orbital.

9.25. Atomic radius trends:

(a) Third period elements. As Z_{eff} increases across the period, the valence electrons are pulled closer and closer to the nucleus. As electrons are added to the 3s and 3p orbitals, electron-electron repulsions eventually become a factor and the rate of decrease falls off.

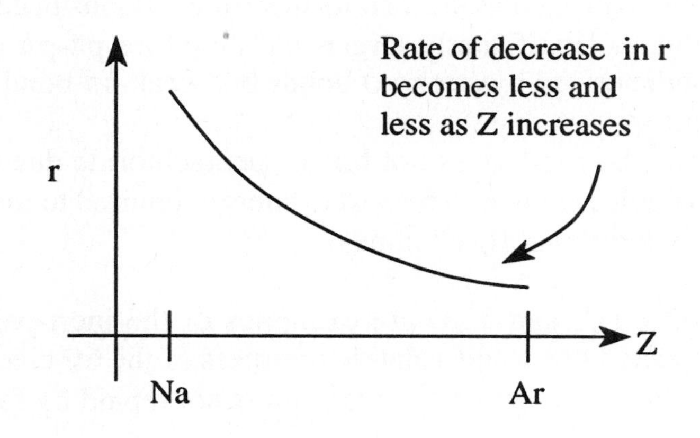

(b) The alkali metals. Their radii increase because the valence electrons occupy the 1s through 7s orbitals that are successively farther from the nucleus. Recall from earlier courses that n, the prinicipal quantum number, is, among other things, indicative of the size of the orbital.

9.27. Mg [Ne]$3s^2$ The ionization energy is less for aluminum because the 3p elec-
Al [Ne]$3s^2 3p^1$ tron is higher in energy than the 3s electron and therefore easier to remove. $E_{3s} < E_{3p}$ because the 3s orbital penetrates through the neon shell better than a 3p orbital.

9.29. Electron affinity is the energy involved when an electron is added to a neutral, gaseous atom (or a gaseous ion). As Z_{eff} increases across a period, an incoming electron will be more attracted and release more energy when it is added to the atom. The magnitude of EA therefore increases across the period. Given that Z_{eff} (as measured when σ equals the number of inner-core electrons) is a constant going down a group, the vertical trend in EA is to decrease because the electrons are being added to orbitals farther and farther away from the constant Z_{eff} and therefore less energy is released.

9.31. Na [Ne]$3s^1$ The electron affinity of magnesium is less than that of sodium
Mg [Ne]$3s^2$ because the electron being added to Mg must be placed in the 3p orbital which is of higher energy. Therefore, less energy will be released when this electron is added.

9.33. Horizontally, Z_{eff} increases and therefore the electronegativity or the ability of an atom in a molecule to attract electrons to itself would also increase. Vertically, Z_{eff} is constant but, since atomic size increases down a group, the incoming electron density is being drawn to positions farther away from the nucleus. Therefore it follows that the ability to attract electrons is diminished going down the group.

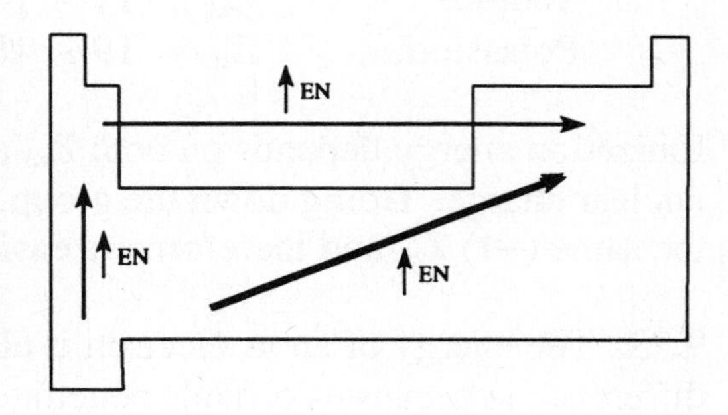

9.35. The units of charge density in Table 9.2 are unit charges per Ångstrom unit or $Å^{-1}$. For example the value for lithium is $+1/0.73Å = 1.4Å^{-1}$.

9.37. The planar molecular layers of graphite are held together by pi bonds formed by the parallel overlap of 2p orbitals. Because silicon atoms are larger than carbon atoms, this $p\pi$-$p\pi$ overlap is much less effective. Silicon is characterized by strong σ bonds but weak $p\pi$ bonds.

9.39. Nitrogen atoms can form strong π bonds because they are so small. Phosphorus, on the other hand, is significantly larger and therefore $p\pi$-$p\pi$ overlap is much less effective. Phosphorus is characterized by strong σ bonds but weak $p\pi$ bonds.

9.41. Bismuth does not form a pentachloride due to the inert pair effect. The $6s^2$ electron pair is difficult to ionize. The cost in energy required to ionize these two electrons is not repaid by forming two additional Bi-Cl bonds.

9.43. TlX and Tl_2O are examples of the inert pair effect in which the valence is two less than expected due to the relative inertness of the $6s^2$ electron pair. The cost in energy required to ionize the two $6s^2$ electrons of thallium is not repaid by forming two additional Tl-X or Tl-O bonds.

9.45. The electronic configurations of Al, Ga, In, and Tl written in groups for the purposes of calculating effective nuclear charges using Slater's rules are as follows:

Al $1s^22s^22p^63s^23p^1$ $Z_{eff} = 13 - 2(0.35) - 8(0.85) - 2(1.00) = 3.50$
Ga $1s^22s^22p^63s^23p^63d^{10}4s^24p^1$ $Z_{eff} = 31 - 2(0.35) - 10(0.85) - 18(1.00) = 3.80$
In $1s^22s^22p^63s^23p^63d^{10}4s^24p^64d^{10}5s^25p^1$ $Z_{eff} = 49 - 2(0.35) - 10(0.85) - 36(1.00) = 3.80$
Tl $1s^22s^22p^63s^23p^63d^{10}4s^24p^64d^{10}4f^{14}5s^25p^65d^{10}6s^26p^1$ $Z_{eff} = 81 - 2(0.35) - 10(0.85) - 68(1.00) = 3.80$

The first ionization energies decrease fairly dramatically going down Group 3A but the effective nuclear charges stay about constant. The reason for the decrease in IE, then, is not Z_{eff}, but rather the increasing distance of the np electron from the effective nuclear charge making it easier to remove this valence electron.

9.47. The diagonal effect states that a special "diagonal relationship" exists between lithium and magnesium, beryllium and aluminum, and boron and silicon. This diagonal connection is due to the fact that lithium, beryllium, and boron have ionic radii, charge densities, and electronegativities more similar to those of magnesium, aluminum, and silicon, respectively, than to the corresponding properties of the elements directly below them in their respective groups. While similarities within a group are still the most important single organizing principle in understanding the periodic table, the diagonal effect means that the Li/Mg, Be/Al, and B/Si pairs are often surprisingly similar to each other.

9.49. (a) Moving horizontally from left to right across the metal-nonmetal line means that the elements move from being metals (Na, Mg, Al) to metalloids (Si) to nonmetals (P, S) and (b) metals (Zn, Ga) to metalloids (Ge, As) to nonmetals (Se, Br).

Chapter 10
Hydrogen and Hydrides

The sections and subsections of this chapter are listed below.

10.1 The Origin of the Elements (and of Us!)
10.2 The Discovery, Preparation, and Uses of Hydrogen
10.3 Isotopes of Hydrogen
10.4 Radioactive Processes involving Hydrogen
  Alpha and Beta Decay, Nuclear Fission, and Deuterium
  Tritium
10.5 Hydrides and the Network
  Covalent Hydrides
  Ionic Hydrides
  Metallic Hydrides
10.6 The Role of Hydrogen in Various Alternative Energy Sources
  Hydrogen Economy
  Nuclear Fusion

Chapter Objectives

You should be able to

- explain the nature of the "big bang" theory
- give a brief general account of the origin of the elements
- represent the nucleus of an element properly
- explain the source of energy in nuclear reactions
- explain the difference between fusion and fission
- relate how and by whom hydrogen was discovered
- explain how hydrogen gas is commonly generated in the laboratory
- describe the common industrial preparations and uses of hydrogen
- describe the dangers and uses of the reaction between hydrogen and oxygen
- describe the similarities and differences among the three isotopes of hydrogen
- describe the isolation and uses of "heavy water," D_2O
- describe, represent, and explain the deuteration of various compounds
- describe and give examples of α, β^-, and β^+ decay
- explain the concept of half life as it applies to radioactive isotopes
- explain and represent fission
- describe the conditions for the occurrence of a nuclear fission chain reaction
- relate the first example of nuclear transmutation
- describe how tritium is produced in nature and by humankind
- discuss the placement of hydrogen among the groups of the representative elements
- describe the three major oxidation states of hydrogen
- describe and give examples of covalent hydrides
- describe and give examples of ionic or "saline" hydrides and their uses
- describe and give examples of metallic hydrides and their uses
- discuss the nature of nonstoichiometric interstitial hydrides
- describe what is meant by the "hydrogen economy"
- describe the potential of nuclear fusion as a future energy source

10.1. $^{11}_{5}B$ 5 protons, 6 neutrons
$^{17}_{8}O$ 8 protons, 9 neutrons
$^{60}_{27}Co$ 27 protons, 33 neutrons
$^{239}_{94}Pu$ 94 protons, 145 neutrons

10.3. (a) $^{12}_{6}C + ^{1}_{1}H \rightarrow ^{13}_{7}N$
(b) $^{13}_{7}N \rightarrow ^{13}_{6}C + ^{0}_{+1}e$ **(positron)**
(c) $^{13}_{6}C + ^{1}_{1}H \rightarrow ^{14}_{7}N$
(d) $^{14}_{7}N + ^{1}_{1}H \rightarrow ^{15}_{8}O$
(e) $^{15}_{8}O \rightarrow ^{15}_{7}N + ^{0}_{+1}e$ **(positron)**
(f) $^{15}_{7}N + ^{1}_{1}H \rightarrow ^{12}_{6}C + ^{4}_{2}He$ **(alpha particle)**

10.5. (a) $^{1}_{1}H + ^{1}_{1}H \rightarrow ^{2}_{1}H + ^{0}_{+1}e$

$$\Delta m = \Sigma m(\text{reactants}) - \Sigma m(\text{products})$$
$$= 2(1.007825) - (2.0140 + 5.488 \times 10^{-4})$$
$$= 0.0011 \text{ u}$$

$$E = (0.0011 \text{ u})\left(\frac{8.984 \times 10^{10} \text{ kJ/mol}}{\text{u}}\right) = 9.88 \times 10^{7} \text{ kJ/mol}$$

(b) proton-proton cycle: $4\,^{1}_{1}H \rightarrow ^{4}_{2}He + 2\,^{0}_{+1}e$

$$\Delta m = \Sigma m(\text{reactants}) - \Sigma m(\text{products})$$
$$= 4(1.007825) - [4.0015 + 2(5.488 \times 10^{-4})]$$
$$= 0.0287 \text{ u}$$

$$E = (0.0287 \text{ u})\left(\frac{8.984 \times 10^{10} \text{ kJ/mol}}{\text{u}}\right) = 2.58 \times 10^{9} \text{ kJ/mol}$$

10.7. While Robert Boyle was, according to Asimov, the first chemist to collect a gas and demonstrate that the volume of a gas is inversely related to its pressure, no mention is made about his work with hydrogen. Even though others (including Boyle) had worked with hydrogen before him, Henry Cavendish is credited with the discovery of this element because, to quote Asimov, he "was the first to investigate its properties systematically."

10.9. $\Delta H = \Delta H^\circ_f[CO(g)] + \Delta H^\circ_f[H_2(g)] - \Delta H^\circ_f[C(\text{graphite})] - \Delta H^\circ_f[H_2O(g)]$
$\Delta H = (-110.5) + 0 - 0 - (-241.8) = 131.3$ kJ/mol

Since this is an endothermic reaction, heat is absorbed and is a reactant. Therefore when the temperature is raised to higher values, this reaction, according to Le Châtelier's principle, shifts to the right. Recall from your previous study of this principle that when the pressure of a reaction mat equilibrium is decreased, it will shift in the direction producing the larger number of moles of gases. This reaction goes from one mole of gaseous reactants to two moles of gaseous products, therefore lowering the pressure will shift the reaction to the right and increase its yield.

(Standard heats of formation vary somewhat from source to source, so your values may not be exactly the same as those given here.)

10.11. $MO(s) + H_2(g) \rightarrow M(s) + H_2O(l)$
M is reduced from +2 to 0; H is oxidized from 0 to +1
$Ag_2O(s) + H_2(g) \rightarrow 2Ag(s) + H_2O(l)$
$Bi_2O_3(s) + 3H_2(g) \rightarrow 2Bi(s) + 3H_2O(l)$

10.13. The atomic weight or mass of an element is the weighted average of a collection of its isotopes. In this case,

$$AW_H = 0.99985(1.007825) + 0.00015(2.0140) = 1.00798\text{u}$$

10.15. Temperature is a measure of mean kinetic energy, which is given by $1/2mv^2$. The heavier D_2O molecules move slower than the lighter H_2Os and therefore need a higher temperature to acheive sufficient velocity to escape from the surface of the solid or the liquid, that is, to melt or boil.

10.17. The electrolysis of water first produces H_2 and O_2 gases. Although naturally occurring water also contains small amounts of D_2O and HDO, these are electrolyzed to D_2 (and O_2) slower because the heavier D^+ cation travels more slowly to the electrodes. The result is that the concentration of D_2O increases in a sample of water as it is being electrolyzed. As D_2O has a greater density than H_2O, the overall density of the water sample increases during the process of electrolysis.

10.19. First we can calculate the grams of D_2O per L of ordinary water.

$$\left(\frac{1 \text{ cup } D_2O}{400 \text{ gal water}}\right)\left(\frac{0.236\text{L } D_2O}{1 \text{ cup } D_2O}\right)\left(\frac{1000\text{mL } D_2O}{\text{L } D_2O}\right)\left(\frac{1.10 \text{ g } D_2O}{\text{mL } D_2O}\right)\left(\frac{1 \text{ gal water}}{3.78\text{L water}}\right) = 0.172 \frac{\text{g } D_2O}{\text{L water}}$$

Next we can calculate the liters of water in the oceans.

$$(3.2 \times 10^8 \text{ km}^3)\left(\frac{1000\text{m}}{\text{km}}\right)^3\left(\frac{100\text{cm}}{\text{m}}\right)^3\left(\frac{1\text{mL}}{1 \text{ cm}^3}\right)\left(\frac{1\text{L}}{1000\text{mL}}\right) = 3.2 \times 10^{20} \text{ L}$$

Finally we are able to calculate the grams of D_2O in the oceans of the world.

$$(3.2 \times 10^{20} \text{ L})\left(0.172 \frac{\text{g } D_2O}{\text{L water}}\right) = 5.5 \times 10^{19} \text{ g } D_2O \text{ (whew!)}$$

10.21. Assume 1 mole of each substance:

$$H_2\left(\frac{1.0080 \times 2 \text{ g}}{\text{mol}}\right)\left(\frac{1 \text{ mol}}{22.4\text{L}}\right) = 0.0900 \text{ g/L} \quad D_2\left(\frac{2.0140 \times 2 \text{ g}}{\text{mol}}\right)\left(\frac{1 \text{ mol}}{22.4\text{L}}\right) = 0.180 \text{ g/L}$$

Assuming ideal behavior, the calculated densities are very close to the actual.

10.23. Only the hydrogen atoms bound directly to oxygen atoms will be replaceable with deuterium when mixed with heavy water, D_2O. This is because the polar O-D bonds of deuterium oxide will interact with the polar O-H bonds of the glucose but not with the essentially nonpolar C-H bonds.

10.25. The allies support an undercover mission to destroy a heavy water plant under Nazi control in Norway. The allies were concerned that the heavy water could be used as a moderator in a possible Nazi effort to produce an atomic bomb.

10.27. beta-minus decay: ${}^{40}_{19}K \xrightarrow{\beta^-} {}^{0}_{-1}e + {}^{40}_{20}Ca$
beta-plus decay: ${}^{40}_{19}K \xrightarrow{\beta^+} {}^{0}_{+1}e + {}^{40}_{18}Ar$

10.29. ${}^{238}_{92}U + {}^{1}_{0}n \rightarrow [{}^{239}_{92}U] \xrightarrow{\beta^-} {}^{0}_{-1}e + {}^{239}_{93}Np$

10.31. ${}^{235}_{92}U + {}^{1}_{0}n \rightarrow {}^{96}_{37}Rb + {}^{138}_{55}Cs + 2{}^{1}_{0}n$

10.33. ${}^{113}_{48}Cd + {}^{1}_{0}n \rightarrow {}^{114}_{48}Cd + \gamma$
${}^{10}_{5}B + {}^{1}_{0}n \rightarrow {}^{4}_{2}He + {}^{7}_{3}Li$

10.35. ${}^{1}_{1}H + {}^{14}_{7}N \xrightarrow{\text{cosmic rays}} {}^{14}_{8}O + {}^{1}_{0}n$

10.37.

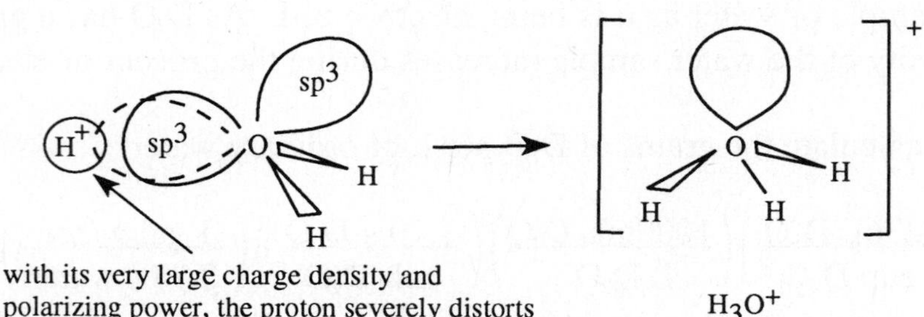

10.39. Using the Born-Haber cycle shown at the right, we see that $\Delta H = 1/2D + EA$

$1/2\ X_2(g) + e^-(g) \xrightarrow{\Delta H} X^-(g)$

$1/2\ X_2(g) \xrightarrow{1/2D} X(g) \xrightarrow{EA} X^-(g)$

For X = H
$\Delta H = 1/2D + EA = 1/2(436.4) + (-77) = 141 kJ/mol$

For X = F
$\Delta H = 1/2D + EA = 1/2(150.6) + (-333) = -258 kJ/mol$

In the process of forming a salt such as NaF, we see that the formation of $F^-(g)$ is an exothermic process and therefore will make a negative (and therefore favorable) contribution to the $\Delta H°_f$ of such alkali metal fluorides. On the other hand if a saline hydride such as NaH is formed, the formation of the corresponding $H^-(g)$ is endothermic and does not make a favorable contribution to the $\Delta H°_f$ of these hydrides. If such hydrides are to form, they must involve highly electropositive metals that have low ionization energies. Such metals will take little energy to ionize and will still have negative lattice energies and therefore have negative standard heats of formation.

10.41. $U_o = 1389 \frac{Z^+Z^-M}{r_o}\left(1 - \frac{1}{n}\right) = 1389 \frac{(+1)(-1)(1.748)}{3.05}\left(1 - \frac{1}{7}\right) = -682\text{kJ/mol}$

where $r_o = r(K^+) + r(H^-) = 1.52 + 1.53 = 3.05$Å; (values from Tables 7.4 and 7.6)
where $n = (9 + 5)/2 = 7$ (values from Table 8.2)

There is a 3.5% difference between the values obtained from the Born-Haber cycle and the Born-Landé equation (with the Born-Haber cycle taken as the most accurate value).

10.43. $U_{Kap} = 1202 \frac{\nu Z^+Z^-}{r_o}\left(1 - \frac{0.345}{r_o}\right) = 1202 \frac{(2)(+1)(-1)}{2.44}\left(1 - \frac{0.345}{2.44}\right) = -846\text{kJ/mol}$

where $r_o = r(Cu^+) + r(H^-) = 0.91 + 1.53 = 2.44$Å

(Note: there is a very small difference, 0.2 on the Pauling scale, between the electronegativities of copper and hydrogen. This result is indicative of a high degree of covalent character in the Cu-H bond. The high degree of covalent character causes the actual (Born-Haber) lattice energy to be much higher in magnitude than that derived from the Kapustinskii equation. See p. 197 for further discussion of these ideas.)

10.45. $U_{Kap} = 1202 \frac{\nu Z^+Z^-}{r_o}\left(1 - \frac{0.345}{r_o}\right) = 1202 \frac{(3)(+2)(-1)}{2.67}\left(1 - \frac{0.345}{2.67}\right) = -2350\text{kJ/mol}$

where $r_o = r(Ca^{2+}) + r(H^-) = 1.14 + 1.53 = 2.67$Å

10.47. $5CaH_2(s) + Ta_2O_5(s) \rightarrow 5CaO(s) + 2Ta(s) + 10H_2(g)$

10.49. Adding these three equations together yields the net reaction $H_2O(g) \rightarrow H_2(g) + 1/2O_2(g)$
The heats of reaction for the three reactions can be calculated as follows:

(1) $\Delta H = \Delta H^\circ_f[H_2(g)] + 6\Delta H^\circ_f[HCl(g)] + \Delta H^\circ_f[Fe_3O_4(s)] - 3\Delta H^\circ_f[FeCl_2(s)] - 4\Delta H^\circ_f[H_2O(g)]$
$= 0 + 6(-92.3) + (-1118.4) - 3(-341.8) - 4(-241.8) = 320.4\text{kJ/mol}$

(2) $\Delta H = 1/2\Delta H^\circ_f[O_2(g)] + 3\Delta H^\circ_f[H_2O(g)] + 3\Delta H^\circ_f[FeCl_3(s)] - \Delta H^\circ_f[Fe_3O_4(s)] - 3/2\Delta H^\circ_f[Cl_2(g)] - 6\Delta H^\circ_f[HCl(g)]$
$= 1/2(0) + 3(-241.8) + 3(-399.5) - (-1118.4) - 3/2(0) - 6(-92.3) = -251.7\text{kJ/mol}$

(3) $\Delta H = 3/2\Delta H^\circ_f[Cl_2(g)] + 3\Delta H^\circ_f[FeCl_2(s)] - 3\Delta H^\circ_f[FeCl_3(s)]$
$= 3/2(0) + 3(-341.8) - 3(-399.5) = 173.1\text{kJ/mol}$

The sum of these heats is $\Delta H_1 + \Delta H_2 + \Delta H_3 = 241.8\text{kJ/mol}$

The heat of the reaction $H_2O(g) \rightarrow H_2(g) + 1/2O_2(g)$ can also be calculated as follows:

$\Delta H = 1/2\Delta H^\circ_f[O_2(g)] + \Delta H^\circ_f[H_2(g)] - \Delta H^\circ_f[H_2O(g)]$
$\Delta H = 1/2(0) + (0) - (-241.8) = 241.8\text{kJ/mol}$

10.51. High temperatures ensure high atomic or nuclear velocities and kinetic energies. This kinetic energy, at the moment of impact of a collision, is largely converted to the potential energy needed to get over the energy of activation of the reaction. High temperatures also increase the number of atomic or nuclear collisions which will serve to increase the rate of these reactions. One significant difference between chemical and nuclear reactions is the magnitude of the energy of activation. The very large repuslsions between nuclei make E_as for nuclear reactions very large indeed.

10.52. ${}^6_3Li + {}^2_1H \rightarrow 2\ {}^4_2He$

Chapter 11
Oxygen, Aqueous Solutions, and the Acid Base Character of Oxides and Hydroxides

The sections and subsections of this chapter are listed below.

11.1 Oxygen
- Discovery
- Occurrence, Preparation, Properties, and Uses

11.2 Water and Aqueous Solutions
- The Structure of the Water Molecule
- Ice and Liquid Water
- Solubility of Substances in Water
- The Self-Ionization of Water

11.3 The Acid-Base Character of Oxides and Hydroxides in Aqueous Solution
- Oxides: Survey and Periodic Trends in
- Oxides in Aqueous Solution (Acidic and Basic Anhydrides)
- The E-O-H Unit in Aqueous Solution
- An Addition to the Network

11.4 The Relative Strengths of Oxo- and Hydroacids in Aqueous Solution
- Oxoacids
- Nomenclature of Oxoacids and Corresponding Salts (Optional)
- Hydroacids

11.5 Ozone

11.6 The Greenhouse Effect

Chapter Objectives

You should be able to

- relate how and by whom oxygen was discovered
- explain why Priestley was able to isolate many more gases than scientists before him
- describe the influence of the phlogiston theory of combustion on the discovery of oxygen
- explain why Priestley could be called the "father of the modern soft drink industry"
- explain the general relationship between carbon dioxide and oxygen in the biosphere
- explain and represent how oxygen gas is generally prepared industrially and in the laboratory
- describe some of the most important uses of oxygen
- describe the structure and polarity of an individual water molecule
- describe the utility of the FONCl rule of hydrogen bonding
- relate the nature and properties of ice and snowflakes to the structure of solid water
- describe and represent the "flickering cluster" model of water structure
- describe, explain, and represent the solubility of ionic and covalent substances in liquid water
- describe and represent the self-ionization of water
- describe the utility and limitations of the concept of the hydronium ion in water and as a product of acids in aqueous solution
- show and explain the distribution of ionic, covalent polymeric, and discrete molecular oxides in the periodic table
- explain why metal oxides are basic anhydrides
- explain why nonmetal oxides are acidic anhydrides
- describe and give an example of an amphoteric anhydride and oxide
- explain why substances containing an M-O-H unit are basic in aqueous solution: give several examples
- explain why substances containing an NM-O-H unit are acidic in aqueous solution; give several examples
- explain the effect on the strength of an oxoacid of changing the central atom; give several examples
- explain the effect on the strength of an oxoacid of changing the number of nonhydroxyl oxygens attached to the central atom; give several examples
- name a variety of oxoacids and their salts
- explain the variation of acid strength in hydroacids within a period and a group of the periodic table; give several examples
- describe and represent ozone
- describe how ozone is formed and some of its beneficial uses
- describe and represent how ozone is formed and destroyed in the stratosphere in the absence of external agents
- describe the greenhouse effect, including its probable causes and how it might be minimized

11.1. $HgO(s) \xrightarrow{heat} Hg(l) + O(g)$

11.3. Of the major components of air (N_2, O_2, H_2O, Ar), only O_2 will react with NO to produce NO_2. Air is approximately 20 percent oxygen while "dephlogisticated air" is, of course, 100 percent oxygen. Therefore, "dephlogisticated air" would have consumed four or five (depending on this experimental accuracy) times as much NO as "common air."

$$NO(g) + 1/2O_2(g) \rightarrow NO_2(g)$$

11.5. ${}^{15}_{8}O \xrightarrow{\beta^+} {}^{0}_{+1}e + {}^{15}_{7}N$

11.7. AW = 0.99763(15.994915) + 0.00037(16.999134) + 0.00200(17.999160) = 15.999u
This result is the same as listed on the periodic table (at least to 3 places to the right of the decimal point).

11.9. The primitive earth had a "reducing atmosphere," that is, there was still much hydrogen present. This kept the oxidation states of the various elements low. As the earth (and life on it) evolved, the oxygen content of the atmosphere rose. This "oxidizing atmosphere" oxidized elements to their higher oxidation states.

11.11. $C_5H_{12}(l) + 8O_2(g) \rightarrow 5CO_2(g) + 6H_2O(l)$

11.13.

(a) NH_3

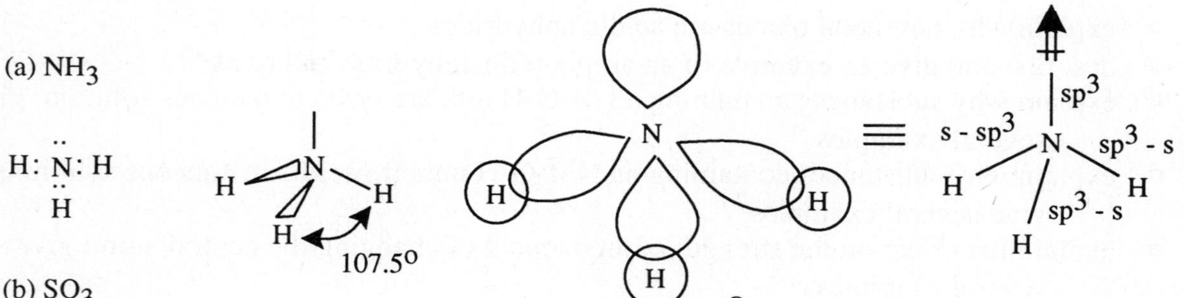

107.5°

s - sp^3 sp^3 - s sp^3 - s sp^3

(b) SO_3

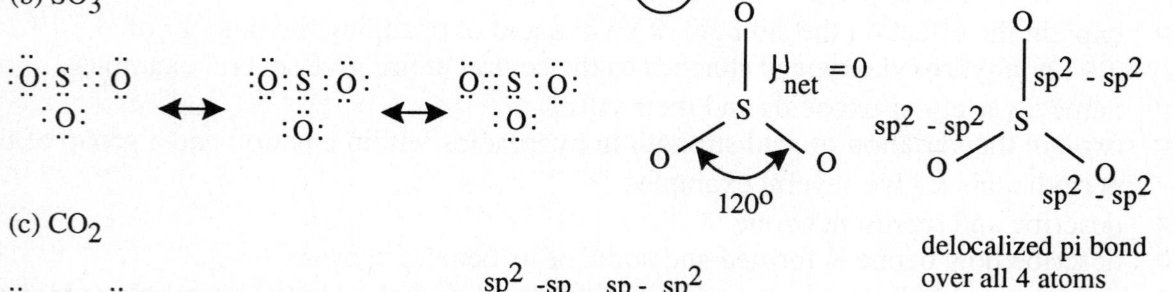

$\mu_{net} = 0$

120°

sp^2 - sp^2 sp^2 - sp^2 sp^2 - sp^2

delocalized pi bond over all 4 atoms

(c) CO_2

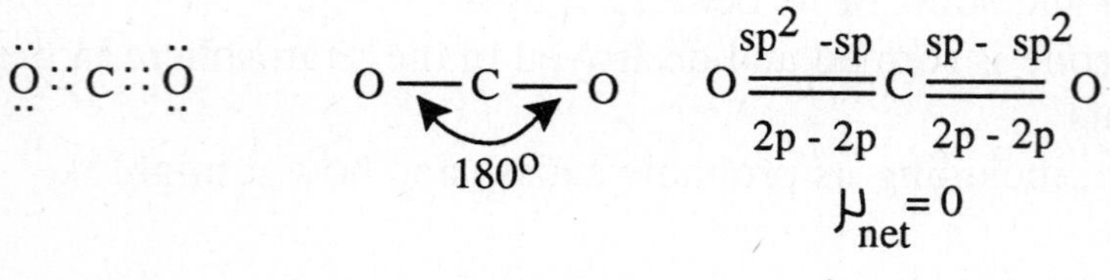

180°

sp^2 -sp sp - sp^2

2p - 2p 2p - 2p

$\mu_{net} = 0$

11.13. (continued)

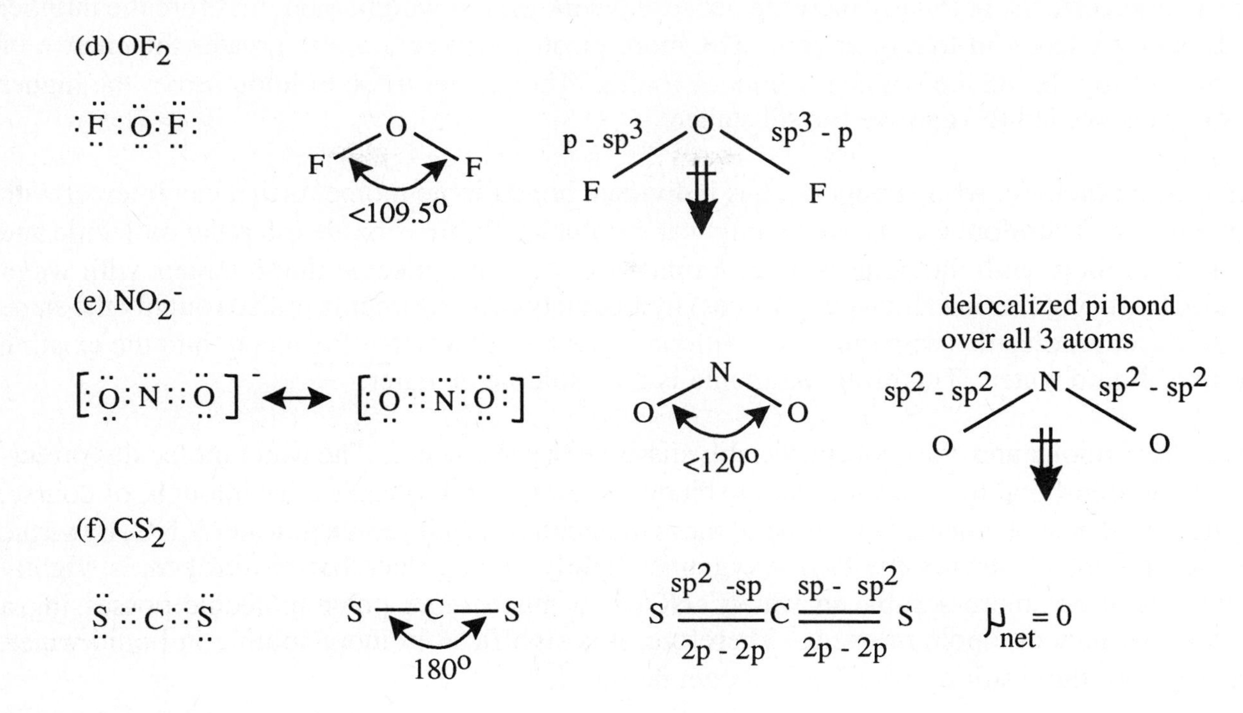

11.15. A lone pair of electrons is confined by (or is under the influence of) only one nucleus whereas a bonding pair is confined by two nuclei. The lone pair is therefore freer to spread out and occupy more space.

11.17. If every zinc and sulfur atom of the wurtzite structure (Figure 7.22d) were replaced by an oxygen of a water molecule (with hydrogen atoms bridging among oxygens), the wurtzite structure would become the ice structure (Figure 11.3).

11.19. Of CH_4, CHF_3, CH_3OH, glucose, and CH_3COOH, only the last three have hydrogen atoms bonded to an electronegative FONCl atom, in this case, oxygen. Only the hydrogen atoms covalently bonded to the oxygen atoms could participate in hydrogen bonds. In CHF_3 the hydrogen atom is not bound to the fluorine atom but rather to the nonFONCl (how about that for a word?) carbon atom.

11.21. Dear English Major:

Ice floats because its structure contains large hexagonally shaped holes. Thus it contains much open space and is light. (A poem could be written about what it would be like to be in one of those open spaces!) When ice melts, water molecules break off the ice structure and wander into some of the holes. Therefore, dear EM, liquid water contains less open space and is heavier than ice. The lighter ice rises to float upon the heavier liquid. (Dear readers: I know you can do better than I did. I would love to see the various short paragraphs that you all come up with. You'll be famous; I might put your paragraph in the next version of this Solutions Manual! (If there is next version.) Submit your entries to the author at Department of Chemistry, Allegheny College, Meadville, PA 16335.

11.23. The boiling points of the Group 7A hydrides (HF, HCl, HBr, HI) should, in the absence of hydrogen-bond effects, gradually increase because the molecular weight (and therefore the number of electrons) increases in that direction. The more electrons there are, the greater the chance of instantaneous dipole-induced dipole (London) forces. The greater these London forces the higher temperature is needed to vaporize the substance.

11.25. Carbon dioxide, while nonpolar, has individual bond dipole moments that can interact with the polar water. Therefore CO_2 is reasonably water-soluble. Sulfur dioxide is a polar molecule and interacts even more with the polar water. Ammonia is a polar molecule that interacts with water molecules by strong (for intermolecular forces) hydrogen bonds. Ammonia is also roughly the same shape as a water molecule (with only an additional hydrogen atom) that fits nicely into the existing structure of liquid water. Therefore, ammonia is very soluble in water.

11.27. O_2 is nonpolar and does not appreciably dissolve in polar water. The water molecules preferentially hydrogen-bond to each other and exclude the dioxygen molecules. (Although, of course, given the abundance of aquatic life, some of these molecules must dissolve in water.) NO, by virtue of the fact that the electronegativity of oxygen is slightly greater than that of nitrogen, is slightly polar and therefore more soluble in water. NO_2 is a much more polar molecule possessing a significant permanent dipole moment. Therefore, it is significantly more soluble in liquid water. The structure of the resulting solution is shown below.

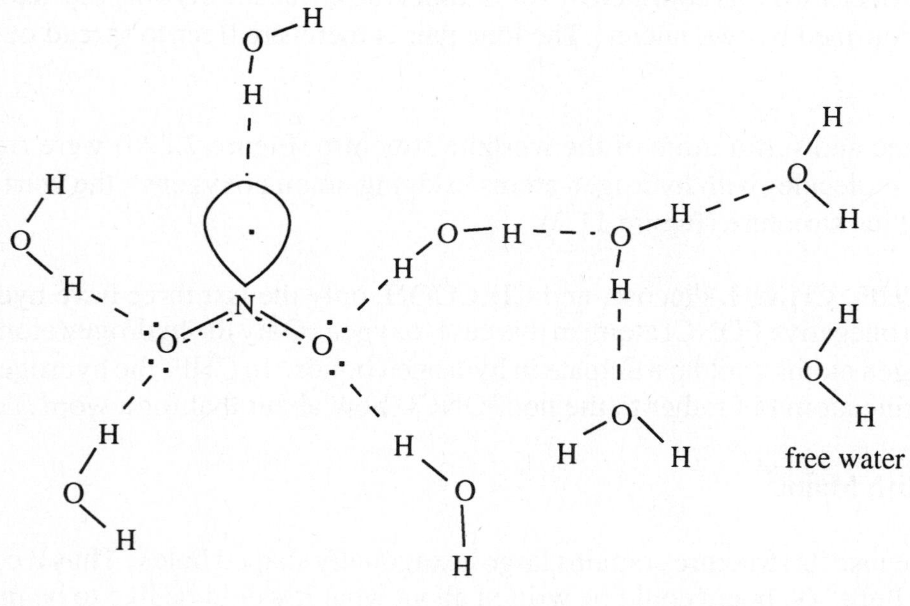

11.29. Glucose has 5 -OH groups and also an oxygen atom in the ring. These 6 polar groupings can interact with water molecules and therefore make glucose soluble. A linear hydrocarbon, on the other hand, is nonpolar. The water molecules will therefore prefer to hydrogen-bond among themselves and exclude (and therefore not dissolve) this solute.

11.31. $H_2O + H_2O \rightleftharpoons H_3O^+(aq) + OH^-(aq)$
$K_w = [H_3O^+(aq)][OH^-(aq)] = x^2 = 1.0 \times 10^{-14}$
$x = [OH^-(aq)] = [H_3O^+(aq)] = 1.0 \times 10^{-7}$
$pH = -\log[H_3O^+(aq)] = 7.00$

11.33. In aqueous solution, the equilibrium among hydronium ions, hydroxide ions, and water molecules must always be satisfied. That is, $K_w = [H_3O^+(aq)][OH^-(aq)]$ must always be equal to 1.0×10^{-14}. Therefore, if $[H_3O^+(aq)]$ is raised, then it follows that $[OH^-(aq)]$ must be lowered such that K_w is still equal to 1.0×10^{-14}.

11.35. Strong acids provide a high concentration of H_3O^+, hydronium ions, in solution. These species are hydrated by other H_2O groups as shown to the right. If these hydrates are stable enough (and of high enough concentrations) they may be isolated from such solutions.

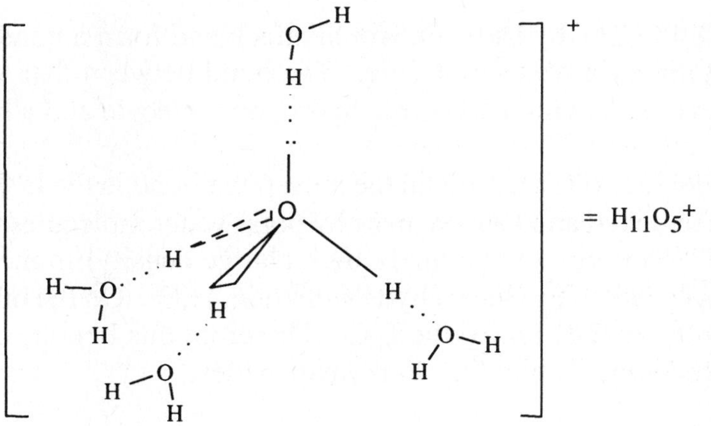

$= H_{11}O_5^+$

11.37. Oxides:

K_2O	Ga_2O_3	As_2O_3 or As_2O_5	SeO_2 or SeO_3
basic	amphoteric (along M/NM border)	acidic but close to borderline	acidic

11.39. The existence of S_3 (if analogous to O_3) depends on the strength of S=S double bonds. Sulfur atoms are considerably larger than oxygen atoms (re: uniqueness principle) and therefore parallel overlap among their p orbitals (so-called pπ-pπ overlap) is much less effective. Therefore, we are unlikely to find an S_3 analog of ozone.

11.41. $N_2O_3 + H_2O \rightarrow 2HNO_2 \rightarrow 2H^+(aq) + NO_2^-(aq)$

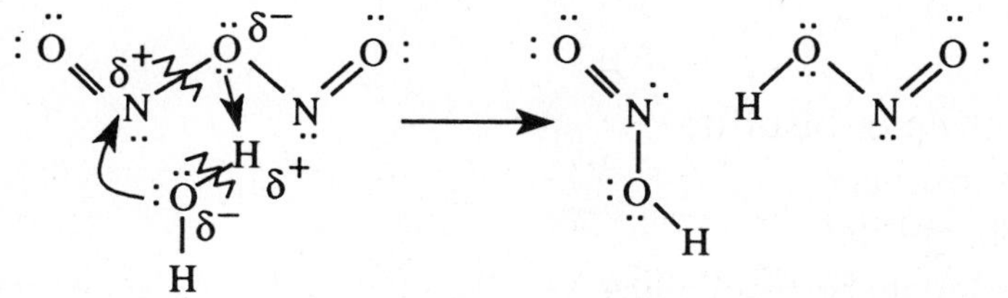

11.43. (One resonance structure of SO_2 is used here for clarity.)

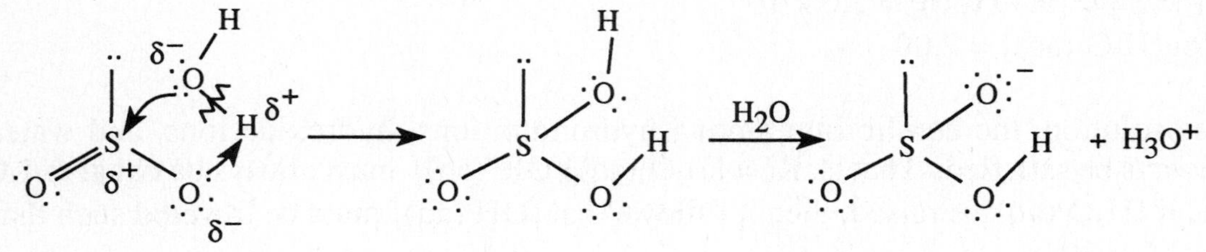

H_2O $+ H_3O^+$ $\delta-$ $\delta+$

11.45. The hydride ion would attack and form a bond with one of the partially positive hydrogen atoms of a water molecule. The bond between that atom and the oxygen of the water would be severed leaving a diatomic hydrogen molecule and a hydroxide ion as products.

11.47. If E is a nonmetal the most polar bond in the E-O-H unit is the O-H. Therefore the O-H bond is attacked and broken apart by polar water molecules to produce EO^- and H^+ ions in solution. The H^+ ion (with its extremely high charge density) immediately forms a covalent bond with a water molecule to produce a hydronium ion, H_3O^+. On the other hand, if E is a metal, the most polar bond in the E-O-H unit is the E-O. Therefore this bond is preferentially attacked and severed by water producing E^+ and OH^- ions (hydroxides).

11.49.
(a) NO(OH) acidic
(b) $Be(OH)_2$ amphoteric
(c) $Ti(OH)_4$ basic
(d) $Si(OH)_4$ acidic

11.51.
(a) $HIO_4 > HIO_3$ due to one more nonhydroxyl oxygen atom withdrawing electron density from the polar O-H bond of the acid making it more susceptible to attack by the polar water molecules of the solvent
(b) $H_3PO_4 > H_3PO_3$ due to one more nonhydroxyl oxygen atom in H_3PO_4
(c) $HSeO_4^- > HSeO_3^-$ due to one more nonhydroxyl oxygen atom in $HSeO_4^-$
(d) $HClO_2 > HClO$ due to the addition of a nonhydroxyl oxygen atom

11.53.
(a) hypobromous acid
(b) periodic acid
(c) arsenious acid
(d) nitrous acid

11.55.
(a) hydrogen sulfite or bisulfite
(b) dihydrogen arsenate
(c) hydrogen phosphite
(d) hydrogen selenate or biselenate
(e) hypobromite

11.57. H_2S is a stronger hydroacid than H_2O because although the O-H bonds are more polar (due to a larger difference in electronegativities), the S-H bonds are longer, weaker, and easier to break to produce hydrogen ions in aqueous solution.

11.59. HBr > H_2Se > PH_3

11.61. What articles did you find? Is the Antarctic hole in the ozone layer still increasing? How prominent now is the Arctic hole? Or did it start to shrink? Are chlorofluorocarbons still thought to be the primary cause of ozone degradation?

Chapter 12
Group 1A: The Alkali Metals

The sections and subsections of this chapter are listed below.

12.1 Discovery and Isolation of the Elements
12.2 Fundamental Properties and the Network
    Hydrides, Oxides, Hydroxides, and Halides
    Application of the Uniqueness Principle and Diagonal Effect
12.3 Reduction Potentials and the Network
12.4 Peroxides and Superoxides
12.5 Reactions and Compounds of Practical Importance
12.6 Selected Topic in Depth: Metal-Ammonia Solutions

Chapter Objectives

You should be able to

- briefly relate how and by whom the alkali metals were discovered
- rationalize the trends in radii, ionization energies, electron affinities, and electronegativities of the group
- describe the general nature of the hydrides, oxides, hydroxides, and halides of the group
- describe and represent the uniqueness of lithium in the group
- describe and represent the diagonal relationship between lithium and magnesium
- define and use the concepts of oxidation, reduction, oxidizing agent, reducing agent, and standard reduction potential
- describe the role of the standard hydrogen electrode in defining standard reduction potentials
- describe and represent the relationship among standard reduction potentials, change in free energy, and the spontaneity of a reaction
- predict the spontaneity of an oxidation-reduction reaction by analyzing its net reduction potential
- explain the trends in the reduction potentials of the alkali metals
- explain why the lighter alkali metals are more reactive than the heavier
- describe and represent the general preparation, structure, and uses of hydrogen peroxide
- describe the occurrence and uses of superoxides
- describe the application of various alkali metals as fertilizers, treatment for bipolar disorders, reducing agents, battery components, and agents for chronometric procedures
- rationalize the properties of alkali/alkaline earth metal liquid ammonia solutions in terms of the solvated electron
- describe and give several examples of electrides and alkalides

12.1.

Cavendish	report of his discovery of hydrogen, 1766
Priestley	published discovery of dephlogisticated air (oxygen), 1774
Lavoisier	characterization of dephlogisticated air as oxygen, late 1770s
Davy	discovery of Group 1A and 2A elements, 1807-1808
Arfvedson	discovery of lithium, 1817
Bunsen	discovery (with Kirchhoff) of cesium and rubidium by spectroscopy, 1860
Mendeleev	formulation of periodic table, 1869

12.3. According to my dictionary, a voltaic pile, named after Alessandro Volta, is a device that produces current electricity, as distinguished from static electricity.

12.5. Melting point (or freezing point) depression is a so-called colligative property, dependent only upon the number of solute particles present in a solution. Adding some $CaCl_2$ provides other particles (ions, in this case) that make it more difficult to freeze out the "solvent," in this case sodium chloride. In order to freeze out the sodium chloride, a lower temperature must be achieved. Turning this argument around means that the melting point of NaCl is also lowered.

12.7. $^{223}_{87}Fr \xrightarrow{\alpha} {}^{4}_{2}He + {}^{219}_{85}At$

$^{223}_{87}Fr \xrightarrow{\beta^-} {}^{0}_{-1}e + {}^{223}_{88}Ra$

12.9. If even sodium hydroxide is completely ionized in aqueous solution, then all of these bases will appear to be of equal strength when water is used as a solvent. In order to distinguish among them, we will need to use a solvent that is less polar and does not interact as strongly with these compounds.

12.11. These iodides can be prepared by combining the appropriate metal hydroxide with hydroiodic acid,

$$MOH(aq) + HI(aq) \rightarrow MI(aq) + H_2O$$

Equal moles of each reactant should be added carefully (these reactions would be highly exothermic). Given the fact that these metal halides are soluble in water, the resulting solutions would have to be evaporated nearly to dryness or until the iodide precipitates out.

12.13. A saltlike (or ionic or saline) hydride, for example, is one that shows significant ionic character. It follows that a "saltlike nitride" would have appreciable ionic character as well.

12.15. When left out in the open, ordinary salt gathers water. In the summertime in areas of high humidity, salt does not pour well because it sticks together. However, while NaCl is somewhat hygroscopic, LiCl is much more so because the very small lithium cation interacts very strongly with water molecules.

12.17. The standard reduction potential of the standard hydrogen electrode is, by convention (that is, by common agreement among the practitioners of the science) taken arbitrarily to be exactly 0.00 volts.

12.19. $Cl_2(g) + 2e^- \rightarrow 2Cl^-(aq)$ $E° = +1.36v$

Cl_2 would be a good oxidizing agent as it has a tendency to be reduced. $\Delta G° = -262kJ$

12.21. (a) The standard reduction potentials of potassium, rubidium, and cesium are similar because there are two opposing trends going down the group starting at potassium. First, the ionization energy is decreasing (but only slowly) making it easier to remove an electron from the metal. This trend would seem to predict that the elements should be easier to oxidize or more difficult to reduce. Therefore, the standard reduction potentials are predicted to become more negative going from potassium to cesium. However, the second trend is in the radii of the resulting cations. These radii slowly increase from potassium to cesium and therefore the charge densities decrease. It follows that these less polarizing cations (K^+ to Cs^+) interact less with water and their solutions are somewhat less stable. [For readers who have covered Chapter 8, the energy of hydration (that is, the energy released when water interacts with a solute, in this case the M^+ cations) is directly proportional to the charge density of the ion. Therefore, it follows that less hydration energy is released going from K^+ to Cs^+ and the oxidation reaction is less favorable for the heavier, larger cations.] It follows that the trend in radii (and charge density) predicts that these elements should be less readily oxidized and more easily reduced from K^+ to Cs^+. Therefore the standard reduction potentials are predicted to become less negative in that direction. It turns out that these two opposing factors balance out fairly equally and the E°'s remain nearly constant for these three elements.

(b) For lithium, however, its unusually small size makes its charge density much larger and the energy of hydration much greater. Another way to say this is that the highly polarizing Li^+ cation strongly interacts with surrounding water molecules and leads to a more stable $Li^+(aq)$ state, a higher tendency to be oxidized, and a lower tendency to be reduced. Thus the standard reduction potential of lithium is significantly more negative than that of its congeners.

12.23. Hydrogen peroxide should be very soluble in water due to the formation of hydrogen bonds among the O-H groups and lone pairs of H_2O_2 and the water molecules.

12.25. Based on the ideas we have discussed under the uniqueness priniciple, the O-O bond distance in H_2O_2 is considerably shorter than the corresponding S-S bond distance in H_2S_2. This longer bond distance lets the lone pairs on the two sulfur atoms rotate by each other freely so that there is no significant hindered rotation in dihydrogen disulfide.

12.27.
$$2\ [H_2O_2(aq) + 2H^+(aq) + 2e^- \rightarrow 2H_2O(l)] \qquad E° = +1.77v$$
$$1\ \underline{[2H_2O(l) \rightarrow O_2(g) + 4H^+(aq) + 4e^-]} \qquad \underline{E° = -1.23v}$$
$$2H_2O_2(aq) \rightarrow O_2(g) + 2H_2O(l) \qquad E° = +0.54v$$

12.29. Half reactions under acid conditions:

$$2\,[Co^{2+}(aq) \rightarrow Co^{3+}(aq) + e^-] \qquad E° = -1.82v$$
$$1\,[2H^+(aq) + H_2O_2(aq) + 2e^- \rightarrow 2H_2O(l)] \qquad E° = +1.77v$$
$$2Co^{2+}(aq) + 2H^+(aq) + H_2O_2(aq) \rightarrow 2Co^{3+}(aq) + 2H_2O(l) \qquad E° = -0.05v$$

Under standard state conditions, H_2O_2 would not oxidize $Co^{2+}(aq)$ to $Co^{3+}(aq)$.

12.31. In order to decide whether sodium metal could be used to reduce Al^{3+} to aluminum metal, we need to calculate the E° of the following equation:

$$Al^{3+}(aq) + Na(s) \rightarrow Al(s) + Na^+(aq)$$

$$1\,[Al^{3+}(aq) + 3e^- \rightarrow Al(s)] \qquad E° = -1.66v$$
$$3\,[Na(s) \rightarrow Na^+(aq) + e^-] \qquad E° = +2.71v$$
$$Al^{3+}(aq) + 3\,Na(s) \rightarrow 3\,Na^+(aq) + Al(s) \qquad E° = +1.05v$$

Yes, sodium could be used to reduce $Al^{3+}(aq)$ to Al(s).

12.33. (a) An analysis of the standard reduction potentials shows that (under standard state conditions), permanganate is capable of oxidizing hydrogen peroxide to diatomic oxygen gas. In this case, hydrogen peroxide is oxidized and acts as the reducing agent.

$$5\,[H_2O_2(aq) \rightarrow O_2(g) + 2H^+(aq) + 2e^-] \qquad E° = -0.68v$$
$$2\,[MnO_4^-(aq) + 8H^+(aq) + 5e^- \rightarrow Mn^{2+}(aq) + 4H_2O] \qquad E° = +1.51v$$
$$5H_2O_2(aq) + 2MnO_4^-(aq) + 6H^+(aq) \rightarrow 5O_2(g) + 8H_2O(l) + 2Mn^{2+}(aq) \qquad E° = +0.83v$$

(b) On the other hand hydrogen peroxide can oxidize $I^-(aq)$ to $I_2(s)$ and therefore serve as an oxidizing agent.

$$1\,[H_2O_2(aq) + 2H^+(aq) + 2e^- \rightarrow 2H_2O(l)] \qquad E° = +1.77v$$
$$1\,[2I^-(aq) \rightarrow I_2(s) + 2e^-] \qquad E° = -0.54v$$
$$H_2O_2(aq) + 2H^+(aq) + 2I^-(aq) \rightarrow 2H_2O(l) + I_2(s) \qquad E° = +1.23v$$

12.35. $4\,KO_2(s) + 2H_2O \rightarrow 4KOH(s) + 3O_2(g)$
K^+ is a spectator ion here. The superoxide (oxidation state = - 1/2) is disproportionating (i.e., is being both oxidized and reduced) with some being oxidized to O_2 (oxidation state = 0) and some being reduced to OH^- (oxidation state of O = -2).

$$1\,[3e^- + 2H^+ + O_2^- \xrightarrow{\text{GER, reduced}} 2OH^-]$$
$$3\,[O_2^- \xrightarrow{\text{LEO, oxidized}} O_2 + 1e^-]$$
$$4O_2^- + 2H^+ \rightarrow 2OH^- + 3O_2$$
$$2OH^- \qquad + 2OH^- \quad \text{(added because of basic conditions)}$$
$$4O_2^- + 2H_2O \rightarrow 4OH^- + 3O_2$$

12.37. Lithium and magnesium are related by the diagonal effect. A regular dose of lithium might keep the magnesium-calcium balance relatively stable. Caution, however, is advised. Much is unknown about this treatment.

12.39. (a) $Cr_2O_3(s) + 6Na(s) \rightarrow 2Cr(s) + 3Na_2O(s)$
(b) $Al_2O_3(s) + 6Na(s) \rightarrow 2Al(s) + 3Na_2O(s)$

12.41.

$$Li(s) \xrightarrow{\text{LEO, oxidized}} Li^+ + e^-$$

$$2e^- + SO_2 \xrightarrow{\text{GER, reduced}} S_2O_4^{2-}$$

Li(s) is losing electrons, is oxidized, and is the reducing agent (LEORA).
$SO_2(g)$ is gaining electrons, is reduced, and is the oxidizing agent (GEROA).

12.43. (a) ${}^{137}_{55}Cs \xrightarrow{\beta^-} {}^{0}_{-1}e + {}^{137}_{56}Ba$
(b) 20 half-lives x 30 yrs/half-life = 600 years or about 6 centuries!

12.45. 12-crown-4, shown at the right, is not quite as large as 18-crown-6 and therefore might not as readily accommodate the large cesium cation.

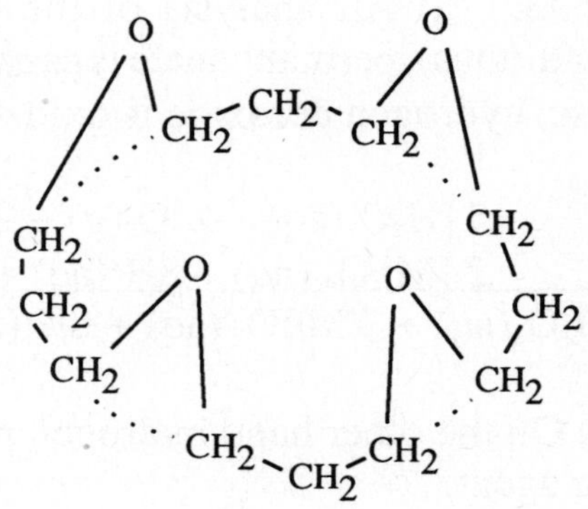

12.47. 2,2,2-crypt is just the right size to bind to the sodium cation. The larger Cs^+ does not fit in the cage of this molecule while the smaller Li^+ slips right through it.

Chapter 13
Group 2A: The Alkaline Earth Metals

The sections and subsections of this chapter are listed below.

13.1 Discovery and Isolation of the Elements
13.2 Fundamental Properties and the Network
- Hydrides, Oxides, Hydroxides, Halides
- Uniqueness of Beryllium and Diagonal Relationship to Aluminum

13.3 Reactions and Compounds of Practical Importance
- Beryllium Disease
- Radiochemical Uses
- Metallurgical Uses
- Fireworks and X-rays
- Hard water

13.4 Selected Topic in Depth: The Commercial Uses of Calcium Compounds
- $CaCO_3$ (limestone)
- CaO (quicklime) and $Ca(OH)_2$ (slaked lime)

Chapter Objectives

You should be able to

- explain the origin of the term "alkaline earth"
- briefly relate how and by whom the alkaline earth metals were discovered
- describe the process by which the Curies isolated radium
- rationalize the trends in radii, ionization energies, electron affinities, and electronegativities of the group
- rationalize, on the basis of standard reduction potentials, why the alkaline earths are good reducing agents
- describe the general nature of the hydrides, oxides, hydroxides, and halides of the group
- list and briefly rationalize the ways in which beryllium is different from its heavier congeners
- list and briefly rationalize the ways in which beryllium is similar to aluminum
- briefly describe the role of the alkaline earths in beryllium disease, neutron production, nuclear fission products, metallurgy, fireworks, x-ray technology, and hard water
- describe how limestone and marble structures are damaged by acid rain
- briefly describe and represent the Solvay process
- briefly describe and represent the use of calcium carbonate in smokestack scrubbers
- briefly list some other uses of calcium carbonate
- briefly describe the use of calcium oxide (quicklime) in the steel and glass industries and in acetylene production
- briefly describe the use of calcium hydroxide (slaked lime) as a mortar, for water treatment, in the paper and pulp industries, and for bleaches
- list several uses of calcium sulfate

13.1. Limelight: a brilliant light created by the oxidation of lime and formerly used in theaters to throw an intense beam of light upon a particular part of the stage. (Limelight is actually the result of heating calcium oxide, CaO, to incandescence with a very hot hydrogen/oxygen or acetylene/oxygen torch. [See J. Chem. Educ., 64 (4), 306 (1987) for further information.]

13.3. $SrCl_2(s) \xrightarrow[\text{electrolysis}]{\text{HCl/KCl}} Sr(s) + Cl_2(g)$

13.5. $3BaO(s) + 2Al(s) \rightarrow Al_2O_3(s) + 3Ba(s)$
$3SrO(s) + 2Al(s) \rightarrow Al_2O_3(s) + 3Sr(s)$

13.7. Epsom salt is $MgSO_4 \cdot 7H_2O$. Magnesia is MgO, a basic anhydride.

$$\underbrace{MgO + H_2O}_{Mg(OH)_2} + H_2SO_4 + 7\,H_2O \rightarrow MgSO_4 \cdot 7H_2O + 2H_2O$$

The net reaction would be as follows: $MgO + H_2SO_4 + 6\,H_2O \rightarrow MgSO_4 \cdot 7H_2O$
Very carefully and slowly add a stoichiometric equivalent of aqueous sulfuric acid to magnesia and then evaporate the solution to dryness.

13.9. We first need to find out what weight 0.01% of seawater is. This is how much magnesium will have to be removed.

$$1.5 \times 10^9\ \text{km}^3 \left(\frac{1000\text{m}}{\text{km}}\right)^3 \left(\frac{100\text{cm}}{\text{m}}\right)^3 \left(\frac{1.025\text{g seawater}}{\text{cm}^3\text{ seawater}}\right)\left(\frac{0.01\text{g Mg}}{100\text{g seawater}}\right) = 1.5 \times 10^{20}\text{g Mg}$$

Now we can calculate how long it would take to consume the above weight of magnesium.

$$1.5 \times 10^{20}\text{g Mg} \left(\frac{1\text{ lb}}{454\text{g}}\right)\left(\frac{1\text{ ton}}{2000\text{ lb}}\right)\left(\frac{\text{yr}}{100 \times 10^6\text{ tons}}\right) = 1.7 \times 10^6\text{ yrs}$$

i.e., between 1 and 2 million years

13.11. $BeF_2(s) + Mg(s) \rightarrow MgF_2(s) + Be(s)$
$\Delta G° = \Delta G°_f\,[MgF_2(s)] + \Delta G°_f\,[Be(s)] - \Delta G°_f\,[BeF_2(s)] - \Delta G°_f\,[Mg(s)]$
$\Delta G° = -1070.2 + 0 - (-979.4) - 0 = -90.8\text{kJ/mol}$

13.13. ${}^{226}_{88}Ra \xrightarrow{\beta^-} {}^{0}_{-1}e + {}^{226}_{89}Ac$

13.15. ${}^{226}_{88}Ra \xrightarrow{\alpha} {}^{4}_{2}He + {}^{222}_{86}Rn$

13.17. Going down the Group 2A elements the radii generally increase as expected. There is not quite as large an increase going from calcium to strontium, as one might expect because the electrons in the filled $3d^{10}$ subshell are not particularly effective at shielding succeeding electrons from the effective nuclear charge. The effective nuclear charge stays approximately constant down the group so that, given the above increase in radii, it gets easier to remove or ionize electrons from these elements. Similarly the ability to attract electrons should fall off as the electrons are added to positions farther away from the Z_{eff}. We see that, in fact, the electronegativities of the Group 2A elements do become smaller going down the group. The electron affinities, however, pose a more difficult problem. The values become less positive (or, if you like, they go more toward the negative) going down the group. This trend means that while it is unfavorable to add an electron to all the alkaline earths, it actually gets a little easier to add electrons to them going down the group. There is no simple explanation of this trend in the alkaline earths.

13.19. The electron affinities of the Group 2A elements are positive because they have filled ns subshells. The electrons to be added, therefore, must be added to np subshells which are of higher energy. Adding electrons to these higher energy orbitals is less favorable (unfavorable in fact) than one would expect on the basis of effective nuclear charges alone.

13.21. We need to see if calcium will reduce water. The half reactions will be as shown below.

$1\ [Ca(s) \rightarrow Ca^{2+}(aq) + 2e^-]$ $E° = +2.87v$; $\Delta G° = -2(96.5kJ/v)(+2.87) = -554kJ$

$\underline{1\ [2H_2O + 2e^- \rightarrow 2OH^-(aq) + H_2(g)]}$ $E° = -0.83v$; $\Delta G° = -2(96.5kJ/v)(-0.83) = +160kJ$

$Ca(s) + 2H_2O \rightarrow Ca^{2+}(aq) + 2OH^-(aq) + H_2(g)$ $\Delta G° = -394\ kJ$

$$E° = -\frac{(-394kJ)}{2(96.5kJ/v)} = 2.04v$$

The reaction in which calcium reduces water is thermodynamically spontaneous under standard state conditions.

13.23.

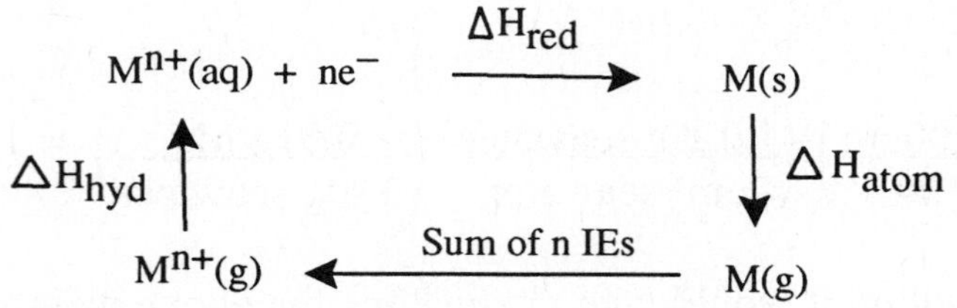

$$\begin{array}{ccc} M^{n+}(aq) + ne^- & \xrightarrow{\Delta H_{red}} & M(s) \\ \uparrow \Delta H_{hyd} & & \downarrow \Delta H_{atom} \\ M^{n+}(g) & \xleftarrow{\text{Sum of n IEs}} & M(g) \end{array}$$

In general, $\Delta H_{red} = -(\Delta H_{atom} + \text{Sum of n IEs} + \Delta H_{hyd})$

$\Delta H_{red}[Li] = -[159 + 520 + (-964)] = 285kJ/mol$ $E° = -3.05v$

$\Delta H_{red}[Be] = -[324 + 899 + 1757 + (-2494)] = -486kJ/mol$ $E° = -1.85v$

As mentioned in the start of Section 13.2, the standard reduction potential for beryllium is not the most negative in its group as is lithium within the alkali metals. The reason for this difference, it was cited in the "Presumably (here we ago again)..." section found on page 328, is "because the energy required to ionize the beryllium to the +2 state is not fully compensated for by the energy released when the Be^{2+} is hydrated." Here we have an opportunity to test the above hypothesis. Note that it takes 2136kJ [= (899+1757) - 520] more to ionize the two electrons of beryllium than the one of lithium. Further note that only 1530kJ [= 2494 - 964] more energy is released when Be^{2+} is hydrated than when Li^+ is. So the above statement comparing the ionization energies with the energies of hydration has been shown to be true.

13.25.

$$^{-}\ddot{H}\quad Ca^{2+}\quad \ddot{H}^{-}\quad 2\ \overset{\delta+}{H}-\overset{\delta-}{\ddot{O}}-H \xrightarrow{H_2O} Ca^{2+}(aq) + H_2\,(g) + 2OH^{-}(aq)$$

13.27. Picture calcium hydroxide as containing two Ca-O-H units as shown below.

$$\overset{\delta+}{H}-\overset{\delta-}{O}-\overset{\delta+}{Ca}-\overset{\delta-}{O}-\overset{\delta+}{H}$$

The most polar bonds in this structure are those between the calcium and the oxygen atoms. The partial charges caused by the larger difference in electronegativity between calcium and oxygen (as opposed to between hydrogen and oxygen) are shown in a larger font. When this compound is placed in water, the polar water molecules will attack the more polar sites within the molecule, that is, the Ca-O bonds. The calcium-oxygen bonds will be broken producing Ca^{2+} and OH^- ions in solution. It follows that this compound will be basic and should be written as $Ca(OH)_2$, calcium hydroxide, rather than H_2CaO_2, which might be referred to as "calcic acid."

13.29. According to my dictionary, formality is defined as "something done merely for form's sake." Referring to a separate Be^{2+} ion is something we do for form's sake. That is, we write this ion in this form because it's similar to that used to write other metallic cations. However, Be(II) compounds are highly covalent due to the high charge density of the Be^{2+} cation so its separate existence as a free ion is highly doubtful. That is, the Be^{2+} is "really a formality."

13.31. $Be(OH)_2 + 2OH^-(aq) \rightarrow Be(OH)_4^{2-}$

$$\left[\begin{array}{c} H \\ :\ddot{O}: \\ H:\ddot{O}:Be:\ddot{O}:H \\ :\ddot{O}: \\ H \end{array}\right]^{2-}$$

All O-Be-O bond angles would be 109.5°; all H-O-Be bond angles would be somewhat less than 109.5°

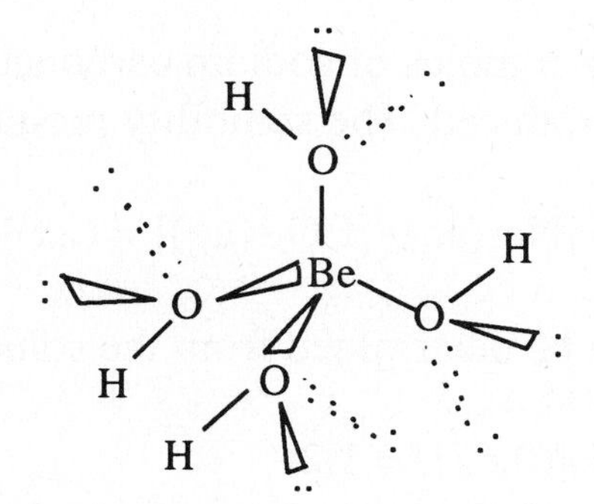

All Be-O bonds would be sp^3-sp^3 ; all O-H bonds would be sp^3 -s

13.33. As we know from the diagonal effect, the aluminum +3 cation will be much like the beryllium +2 cation. Both ions have large charge densities and will form more covalent compounds than we might otherwise expect. Therefore, we might predict that $AlCl_3$ will not be as soluble in water as we would expect. (It turns out, however, that $AlCl_3$ is quite water-soluble.) We also would predict that it will be more soluble in nonpolar organic solvents than otherwise might be expected.

This last prediction is, in fact, quite true as this compound is soluble in carbon tetrachloride and ether. Being electron deficient, aluminium chloride will act as an electron-pair acceptor, that is, as a Lewis acid.

13.35. $^{131}_{53}I \xrightarrow{\beta^-} {}^{0}_{-1}e + {}^{131}_{54}Xe$
While Sr 90 would replace calcium in bone structure and cause radiation damage, I 131 would go to the thyroid where it would be used to make thyroxine.

13.37. $ZrCl_4 + 2Mg(s) \xrightarrow{\Delta} Zr(s) + 2MgCl_2$
$HfCl_4 + 2Mg(s) \xrightarrow{\Delta} Hf(s) + 2MgCl_2$

13.39. $CaCO_3(s) + 2HCl(aq) \longrightarrow H_2O + CO_2(g) + CaCl_2(aq)$

13.41.

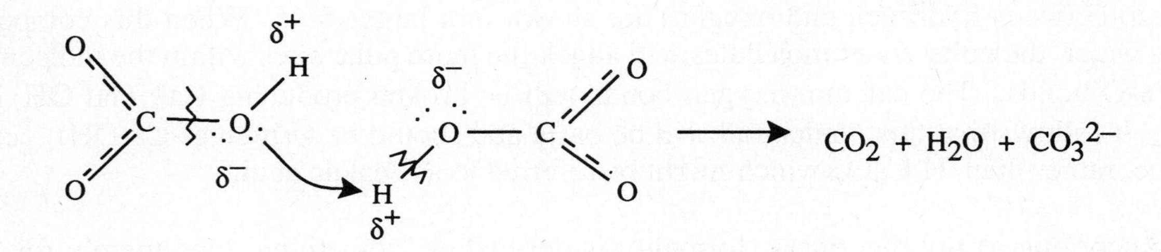

13.43. $CaCO_3(s) + 2HCl(aq) \rightarrow CO_2(g) + H_2O(l) + CaCl_2(aq)$

13.45. $NaHCO_3(aq) + HCl(aq) \rightarrow H_2O + CO_2(g) + NaCl(aq)$

13.47. $Na_2CO_3(s) \rightleftharpoons 2Na^+(aq) + CO_3^{2-}(aq)$

When x moles of sodium carbonate dissolve in 1L of water, 2x moles of Na^+ and x moles of CO_3^{2-} are produced. The solubility product, K_{sp}, will be as follows.

$K_{sp} = [Na^+(aq)]^2[CO_3^{2-}(aq)] = (2x)^2(x) = 4x^3$

x can be determined from the solubility: $x = \left(\frac{7.1g\ Na_2CO_3}{100cm^3}\right)\left(\frac{mol}{106.0g}\right)\left(\frac{cm^3}{mL}\right)\left(\frac{1000mL}{L}\right) = 0.67M$

$K_{sp} = 4(0.67)^3 = 1.2$

Chapter 14
The Group 3A Elements

The sections and subsections of this chapter are listed below.

14.1 Discovery and Isolation of the Elements
- Boron
- Aluminum
- Gallium
- Indium and Thallium

14.2 Fundamental Properties and the Network
- Hydrides, Oxides, Hydroxides, and Halides

14.3 Structural Aspects of Boron Chemistry
- Allotropes
- Borides
- Borates

14.4 Aluminum, Gallium, Indium, and Thallium
- Aluminum Metal and Alloys
- Alums
- Alumina
- Gallium, Indium, and Thallium Compounds

14.5 Selected Topic in Depth: Electron-Deficient Compounds

Chapter Objectives

You should be able to

- briefly state how and by whom the Group 3A elements were discovered
- rationalize the trends in radii, ionization energies, electron affinities, and electronegativities of the group with particular emphasis on the anomalies in these properties shown by the heavier elements
- describe the general nature of the group oxides, hydroxides, and halides
- list and briefly rationalize the ways in which boron is different from its heavier congeners
- list and briefly rationalize how the chemistry of the heavier Group 3A congeners is organized by the inert pair effect
- describe the relationship between the general chemical nature of the borides and their applications
- describe the general molecular nature and a few of the applications of ortho-, meta-, and perborates
- list and briefly rationalize several applications of aluminum and its alloys
- list and briefly rationalize several applications of alums and alumina
- describe the safety and health hazards that accompanied the discovery of the borohydrides
- describe the nature of three-center, two-electron bonds
- draw and rationalize semi-topological diagrams of the simple boranes
- represent and rationalize the preparations and major reactions of the boranes
- describe and represent several electron-deficient compounds other than the boranes

14.1. What were you able to come up with for this one? I would think that a prime candidate for the Latin or Greek basis of the group name would concentrate on the electron deficiency of many of the compounds of these elements. Or perhaps the structure and usefulness of boron and aluminum could be used as the source of a root name. As in other places in this manual, I invite you to submit any ideas you have come up with to me at the Department of Chemistry, Allegheny College, Meadville, PA 16335.

14.3. The sixth Group 3A element would have atomic number 113. The predominant valence of this metal would most likely be +1, it would be of high density (if enough of it could be prepared to measure its density), its oxide (M_2O) would be a basic anhydride, and probably only one even remotely stable isotope could be prepared.

14.5. The AlF_6^{3-} anions are in a face-centered cubic array with the sodium cations in all of the tetrahedral and octahedral holes. There are 4 [= 8(1/8) + 6(1/2)] anions and 12 [= 12(1/4) + 1 + 8] cations in the structure yielding a stoichiometry of Na_3AlF_6.

14.7. ${}^{6}_{3}Li + {}^{1}_{0}n \rightarrow {}^{4}_{2}He + {}^{3}_{1}H$

14.9. We would expect the ionization energies to decrease down the group because the electrons to be ionized are farther and farther away from an approximately constant effective nuclear charge. The actual values do show a general decrease but there are two slight exceptions at gallium and thallium. The rationale for the exceptions has to do with the inability of electrons in filled d and f subshells to shield succeeding electrons from the nuclear charge as effectively as other electrons do. The ns electrons of gallium, indium, and thallium penetrate through these filled subshells and therefore experience a slightly greater effective nuclear charge than would normally be expected.

14.11. We would expect the electronegativities to decrease down the group because the ability of an atom in a molecule to attract electrons to itself will decrease as those electrons are farther and farther away from the effective nuclear charge. However, we find that the electronegativities show a slight increase toward the bottom of the group because these incoming electrons feel a greater than expected effective nuclear charge. The Z_{eff} is greater because the 4p, 5p, and 6p orbitals of gallium, indium, and thallium, respectively, penetrate through the inner shell electrons (of the transition and lanthanide elements) to the nucleus more effectively than otherwise might be expected.

14.13. Boric acid is best thought of as a Lewis acid, accepting the electron pair of an entering hydroxide ion, OH^-, to form the tetrahydroxoborate ion, $B(OH)^-_4$. Recall that an oxoacid is an acid because its E-O-H unit is more polar at the O-H bond. This is not clearly the case in boric acid. The electronegativities of boron and hydrogen are very similar ($EN_B = 2.0$; $EN_H = 2.1$) and therefore boric acid is a very weak acid at best if viewed using the traditional Arrhenius or Brønsted-Lowry definitions.

14.15. Adding a stoichiometric amount of hydrochloric acid to a thallium(I) salt such as the nitrate will precipitate the relatively insoluble thallium(I) chloride:

$$TlNO_3(aq) + HCl(aq) \rightarrow TlCl(s) + HNO_3(aq)$$

In order to prepare thallium(III) salts such as $TlCl_3$, a strong oxidizing agent is required. Chlorine gas itself is strong enough to oxidize the metal to the trichloride.

$$2Tl(s) + 3Cl_2(g) \rightarrow 2TlCl_3(s)$$

14.17. In thallium, the two 6s electrons are difficult to ionize because the filled nd (n = 3, 4, and 5) and 4f subshells do not shield them particularly well from the nucleus. In addition, the thallium atom is large enough that Tl-OH bonds are comparatively long and weak. The result of these two factors involved in the diagonal effect is that the larger than expected second and third ionization energies are not compensated for by the formation of the second and third Tl-OH bonds. Consequently, only TlOH forms instead of $Tl(OH)_3$. In aluminum, the ionization energies are recovered when the two additional, relatively strong Al-OH bonds are formed.

14.19.

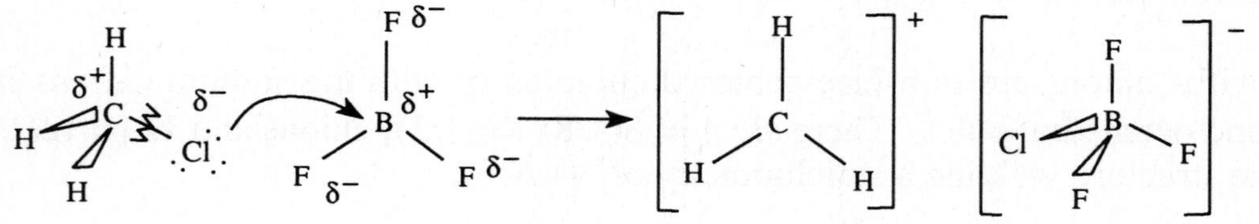

H δ+ C H H, δ− Cl, B δ+ F δ−, F δ−, F δ− ⟶ $[CH_3]^+$ $[BF_3Cl]^-$

14.21. Mechanism of reaction between BX_3 and H_2O:

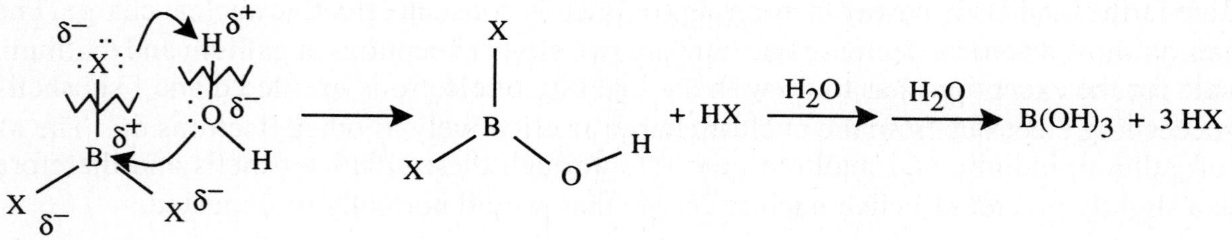

$$BX_3 + H_2O \longrightarrow X_2B\text{–}OH + HX \xrightarrow{H_2O} \xrightarrow{H_2O} B(OH)_3 + 3\,HX$$

14.23. BH_4^- and CH_4 are isoelectronic, that is, they have the same number of electrons. Both are tetrahedral structures in which the central atom is sp^3 hybridized.

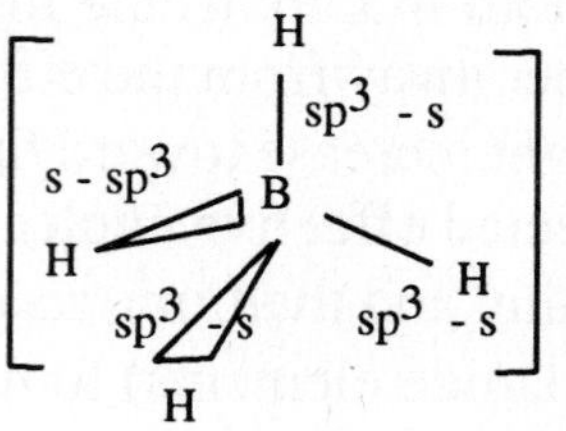

$[BH_4]^-$: B–H bonds, sp^3 - s, s - sp^3, sp^3 - s, sp^3 - s

14.25.

$$Tl^{3+}(aq) + 2e^- \rightarrow Tl^+(aq) \qquad E° = +1.25v;\ \Delta G° = -2(96.5)(1.25) = -241\text{ kJ}$$

$$Tl^+(aq) + e^- \rightarrow Tl(s) \qquad E° = -0.33v;\ \Delta G° = -1(96.5)(-0.33) = +32\text{ kJ}$$

$$Tl^{3+}(aq) + 3e^- \rightarrow Tl(aq) \qquad \Delta G° = -209\text{kJ}$$

$$E° = \frac{-209\text{ kJ}}{-3(96.5\text{kJ/v})} = 0.72v$$

14.27. Eight unit cells are shown in Figure 14.11. There is 1 [= 8(1/8)] Ca^{2+} in the unit cell as well as 1 B_6^{2-} anion located in the center.

14.29. The metal borides are inert, hard, and high-melting because the boride units, while forming regular arrays in a lattice, are bound to each other by a network of B-B bonds. Therefore, when these compounds are to be broken apart in some way, a massive array of interwoven B-B bonds must be severed. For this reason, the metal borides are very difficult to react, crush, or melt.

14.31.

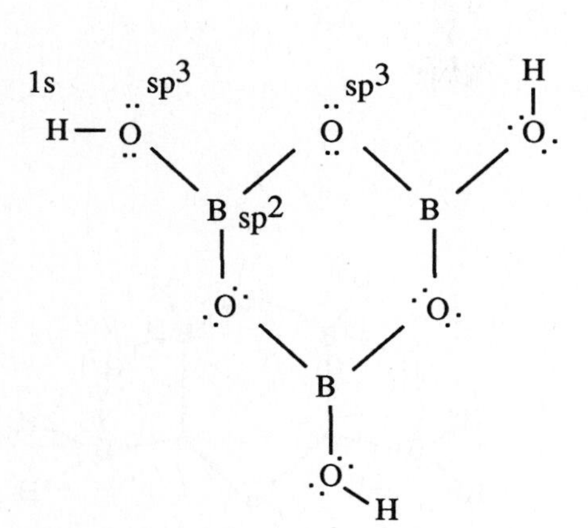

The B-O-H and B-O-B bond angles are somewhat less than 109.5° while the O-B-O angles are 120°.

14.33. All the atoms in $B(OH)_4^-$ are sp^3 hybridized. The O-B-O bond angles are 109.5° while the H-O-B angles are somewhat less than 109.5°.

14.35. Aluminum is resistant to reaction with air and water because it forms a hard, tough oxide film on its exterior surfaces.

14.37. $2Al(s) + 6H^+(aq) \rightarrow 2Al^{3+}(aq) + 3H_2(g)$

$2\ [Al(s) \rightarrow Al^{3+}(aq) + 3e^-]$	E° = +1.66v	ΔG° = -(6)(96.5kJ/v)(1.66v) = -961kJ
$3\ [2H^+(aq) + 2e^- \rightarrow H_2(g)]$	E° = 0.00v	ΔG° = 0
$2Al(s) + 6H^+(aq) \rightarrow 2Al^{3+}(aq) + 3H_2(g)$		ΔG° = -961kJ

14.39. If the $Al(OH)_3$ is allowed to sit in an open container it may dehydrate to produce some Al_2O_3 which is insoluble in water.

14.41. How did your survey come out? Aluminum chlorhydrate is certainly not the only active ingredient in antiperspirant products.

14.43. A "multicenter bond" is one in which two electrons are spread out amongst and serve to bond together more than two atoms. In diborane, B_2H_6, there are two three-center, two-electron bonds, one of which is highlighted in the diagram at the right. One boron and the hydrogen each contribute an electron to the 3c-2e bond but the second boron contributes to the second B-H-B bond and not to the original. Thus there are only two electrons available to hold the three atoms together.

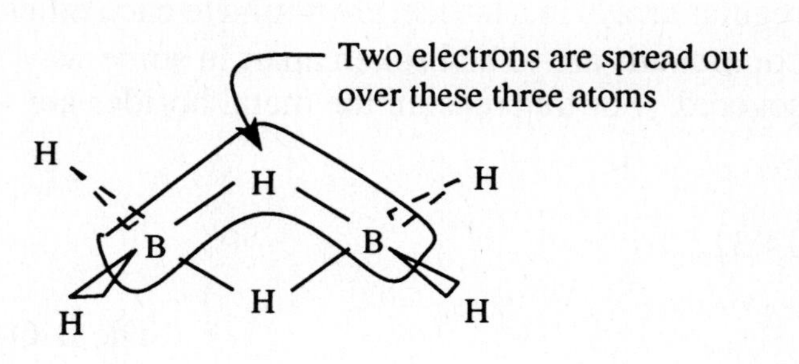

14.45. B_5H_{11} -- consult Figure 14.15(d) for structure

Available electrons:		Distribution of electrons:	
5B x $3e^-$ =	$15e^-$	8 B-H =	$16e^-$
11H x $1e^-$ =	$11e^-$	3 B-H-B =	$6e^-$
	$26e^-$	2 BBB =	$4e^-$
			$26e^-$

14.47. In $B_{12}H_{12}^{2-}$, there are 50 [= (3x12) + (12x1) + 2] electrons which is the same number as there are in $C_2B_{10}H_{12}$. That is, these two structures are isoelectronic. It therefore does seem reasonable that $C_2B_{10}H_{12}$ should exist and have the same basic structure as shown for $B_{12}H_{12}^{2-}$. A large number of geometric isomers (at least 9) could potentially exist as the two carbons could be spread out in various positions in the icosahedron.

14.49. In addition to the "end-to-end" isomer shown in Table 14.6, the "side-to-side" and "end-to-side" structures (as represented schematically at the right) should also be possible.

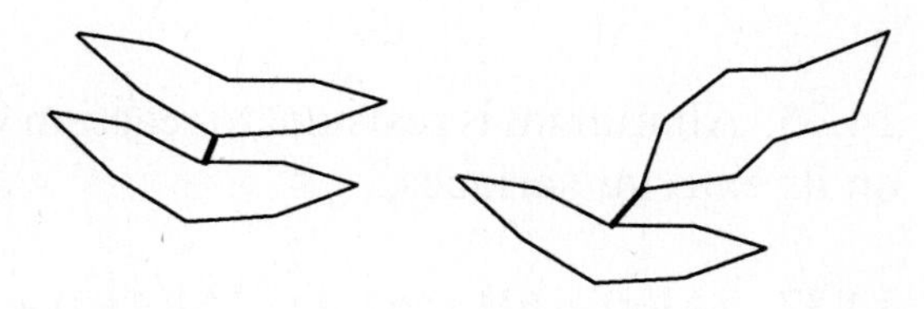

14.51.

14.53. The carbon and boron atoms use sp^3 hybrids. The interior H-B-H bond angles are probably close to 95° as in diborane. The exterior H (or CH_3)-B-H angles are probably greater than 109.5°.

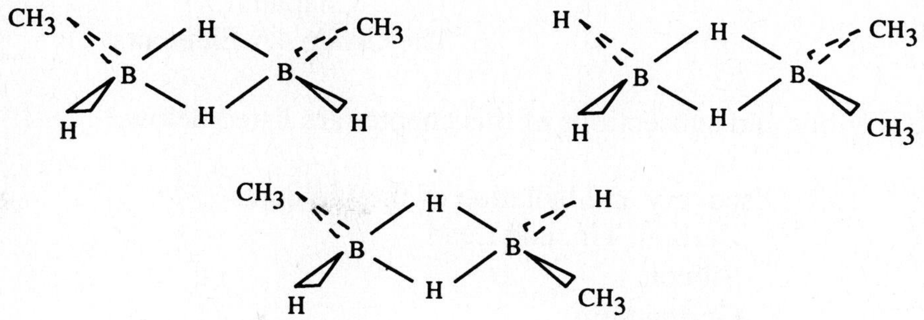

Chapter 15
The Group 4A Elements

The sections and subsections of this chapter are listed below.

15.1 Discovery and Isolation of the Elements
- Carbon, Tin, and Lead
- Silicon
- Germanium

15.2 Fundamental Properties and the Network
- Hydrides
- Oxides and Hydroxides
- Halides

15.3 Reactions and Compounds of Practical Importance
- Graphite and Diamond
- Tin Disease
- Radiochemical Uses
- Carbon Compounds
- Lead Compounds and Toxicology

15.4 Silicates, Silica, and Aluminosilicates
- Silicates and Silica
- Aluminosilicates

15.5 Selected Topics in Depth: Semiconductors and Glass
- Semiconductors
- Glass

Chapter Objectives

You should be able to

- briefly describe the form in which some of the Group 4A elements were known to the ancients
- briefly relate how and by whom the remaining Group 4A elements were discovered
- outline the general manner in which the periodic law, uniqueness principle, inert pair effect, and the acid-base character of oxides are applicable to this group
- explain why catenation is much more prevalent in the carbon hydrides than it is in those of the heavier congeners, particularly silicon
- compare and contrast the incidence of $p\pi$-$p\pi$ bonding in carbon compounds with $p\pi$-$d\pi$ bonding in those of silicon
- describe the general nature of the group oxides, hydroxides, and halides
- compare and contrast the major allotropes of carbon
- briefly describe the role of the Group 4A elements in tin disease, radiochemical chronometry, and tracing methods
- describe several applications of carbon oxides; ionic, covalent, and interstitial carbides; carbon disulfide; and cyanides
- discuss some major applications of lead compounds including the storage battery
- discuss the general toxicology of lead compounds
- list, represent, and discuss the variety of silicates including the relationship of submicroscopic structures to macroscopic properties
- describe the structure and uses of aluminosilicates
- describe the nature of intrinsic, n-type, and p-type Group 4A semiconductors
- represent and describe the preparation of semiconductor-grade silicon
- describe the general nature of common, quartz, laboratory-grade, crystal, colored, photochromic, and fiber glasses

15.1. SnO_2(cassiterite) + 2C(s) → Sn(l) + 2CO(g)

15.3. What important distinctions did you find? Again, I would be most interested in the results of your interview. In any event, I hope it was beneficial for both you and the faculty member(s) you talked with. Inorganic and organic chemists are not so different afterall.

15.5. SiF_4(g) + 4K(l) → Si(s) + 4KF(s), the same process used by Gay-Lussac and Thénard but Berzelius more carefully purified the products.

15.7. The -ium suffix is consistent with the more metallic character of Ge while the -on ending of silicon is consistent with its placement (shown at the right) on the nonmetal side of the metal/ nonmetal line.

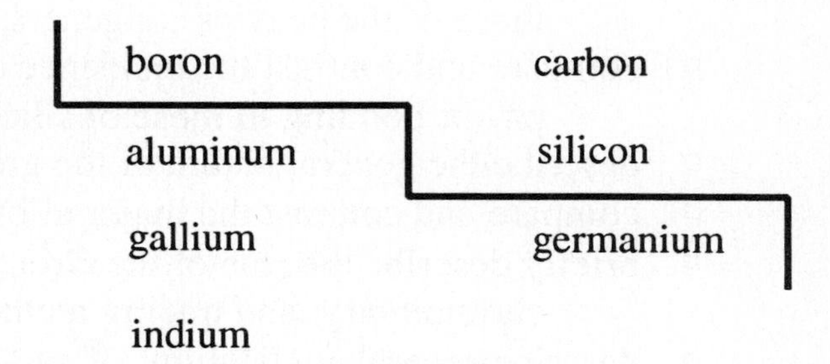

15.9. The Pauling electronegativity values generally decrease going down the group as expected. However, the values for tin and lead are a little higher due to a somewhat greater effective nuclear charge experienced by the incoming bonding electrons.

15.11. Generally, the values of the electron affinities of the Group 4A elements show that it becomes a little more difficult to add electrons to the elements as we go down the group. That is, the values become a little less negative. The exception, as usual, comes with the heavier elements, particularly germanium and tin in this case. In tin, the electron affinity becomes a little more negative than the value for germanium. This is another example of the somewhat increased effective nuclear charge felt by incoming electrons as they are added to orbitals that are not as quite as effectively shielded from the nuclear charge by lower lying d and f subshells.

15.13. At the top of Group 4A is the almost fully covalent methane, CH_4. Going down the group we encounter more and more ionic character in the hydrides.

15.15. "Silons" are unlikely creatures. Presumably they are based on the existence of catenated silane-based molecules but silicon does not catenate very well due to the weakness of Si-Si bonds. The "Horta" (Spock discovered through a mind-meld that this was the name the creature called himself) should not have been as much of a threat as he appeared to be in the *Star Trek* episode. He certainly could not invade the earth because he would readily be oxidized in the earth's oxygen-rich atmosphere. (Question: was the *Star Trek* crew wearing protection against the presumably reducing environment in the mining tunnels of Planet Janus VI where the creature was found?)

15.17. The silicon atom in silane, SiH_4, is larger and contains readily available d orbitals. Si-H bonds are longer and weaker than C-H bonds. When Si-Cl bonds are formed they are more polar than C-Cl bonds. (ΔEN for Si-Cl is 1.2 whereas ΔEN for C-Cl is 0.5.) The additional ionic character of Si-Cl bonds most likely adds to their strength. The most important consideration is probably the available d orbitals of silicon. An attacking chlorine molecule can donate a pair of electrons to the silicon atom (through one of its empty d orbitals) and readily form a low energy transition state. The energy of activation, then, of the SiH_4/Cl_2 reaction is lower than that of the CH_4/Cl_2 reaction.

15.19. The greater size of the silicon atom provides more room (less steric hindrance) for the attack of an incoming molecule and would tend to make such collisions more effective.

15.21. As the covalent radii increase going down the group, the element-to-element bonds get longer and weaker. Therefore, the stability of the chains decreases down the group.

15.23. Ionic character of Group 4A oxides

CO_2	SiO_2	GeO_2	SnO_2	PbO (particularly ionic due to Pb^{II} vs Pb^{IV})
(acid)	(acid)	(ampho)	(ampho)	(ampho)

$\longrightarrow$

increasing ionic character
reflected in acid-base characteristics

15.25. LiCl ionic but more covalent than expected

$BeCl_2$ mostly covalent and characterized by bridging Be-Cl-Be bonds producing a linear polymer

BCl_3 covalent, electron-deficient compound

CCl_4 covalent, saturated, nonpolar compound

15.27. The direct combination of elemental germanium and tin with chlorine yields the tetrachlorides. Other measures must be taken to produce the dichlorides.

(a) $GeCl_4 + Ge \rightarrow 2GeCl_2$
$Ge + 2Cl_2 \rightarrow GeCl_4$

(b) $Sn + HCl(g) \rightarrow SnCl_2 + H_2$
$Sn + 2Cl_2 \rightarrow SnCl_4$

15.29. The carbon halides are not electron deficient as are most of the Group 3A halides. Therefore, the carbon halides need not form bridged structures in order to achieve an octet of electrons around the central atom.

15.31. $PbCl_2$ would be more soluble as it is the more ionic. The charge density of Pb^{2+} is less than that of Pb^{4+} giving the latter a greater polarizing power and an increased ability to form more covalent compounds.

15.33.

$$1\,[2Cl^-(aq) \rightarrow Cl_2(g) + 2e^-] \qquad E° = -1.36v;\ \Delta G° = -(2)(96.5kJ/v)(-1.36v) = 262kJ$$

$$1\,[PbO_2(s) + 4H^+(aq) + 2e^- \rightarrow Pb^{2+}(aq) + 2H_2O(l)] \qquad E° = +1.46v;\ \Delta G° = -(2)(96.5kJ/v)(1.46v) = -282kJ$$

$$2Cl^-(aq) + PbO_2(s) + 4H^+(aq) \rightarrow Cl_2(g) + Pb^{2+}(aq) + 2H_2O \qquad E° = +0.10v;\ \Delta G° = -20kJ$$

This oxidation is possible under standard state conditions.

15.35. No, we would not expect the electrical conductance of diamond to be high. All of the electrons of diamond are tied up in forming the very strong covalent bonds among carbon atoms in this interconnected lattice. In the language of band theory, most all of the electrons in diamond lie in the valence band and very few in the conduction band.

15.37. One reason is that rocks do not take in carbon. Therefore there is no steady state concentration or ratio of C 14 to C 12 in rocks. The second reason is that rocks are generally too old to be measured by C 14 dating. The half-life of C 14 is only about 5730 years, not a very long time compared to the age of rocks which is often in the millions of years.

15.39. Now we can see that radium will be found with uranium of which it is a radioactive daughter and not necessarily with its congener barium.

15.41. Each source of lead is the product of a different combination of three radioactive series. The lead impurities derived from the combustion of coal will have a different set of isotopes than that set from a smelter. A set of isotopes is, in essence, the signature of that particular source of lead. The set of lead isotopes used in a smelter in Missouri is very likely to have a different signature than that of the lead used in a Californian smelter.

15.43.

S::C::S   S—C—S (180°)   S = C = S (sp^2 -sp, sp - sp^2; 2p - 2p, 2p - 2p)

15.45.

H:C:::N: H—C≡N (180°)   H—C≡N (1s - sp; sp - sp, $2p_x - 2p_x$, $2p_y - 2p_y$)

15.47.

$$Pb(s) + SO_4^{2-}(aq) \rightarrow PbSO_4(s) + 2e^- \qquad E° = +0.31v$$

$$PbO_2(s) + SO_4^{2-}(aq) + 4H^+(aq) + 2e^- \rightarrow PbSO_4(s) + 2H_2O \qquad E° = +1.70v$$

$$Pb(s) + PbO_2(s) + 4H^+(aq) + 2SO_4^{2-}(aq) \rightarrow 2PbSO_4(s) + 2H_2O(l) \qquad E° = +2.01v$$

$$\Delta G° = -2(96.5kJ/v)(2.01v) = -388kJ/mol$$

Equation 15.16(c), shown below, is not quite the same as the one above.

$$Pb(s) + PbO_2(s) + 2H^+(aq) + 2HSO_4^-(aq) \rightarrow 2PbSO_4(s) + 2H_2O(l) \qquad 15.16(c)$$

The association of $2H^+(aq)$ and $2SO_4^{2-}(aq)$ into $2HSO_4^-(aq)$ to yield Equation 15.16(c) would decrease the magnitude of $\Delta G°$ about 5 to 6 percent.

15.49. The AlP pair is isoelectronic with two silicon atoms. If we reformulate $AlPO_4$ as AlO_2PO_2 we see that $AlPO_4$ is isoelectronic with two SiO_2 units. Another way to look at this situation is that $AlPO_4$ is SiO_2 (quartz) with half of the silicon atoms replaced by aluminum and the other half replaced by phosphorus. It stands to reason then that aluminum phosphate forms in quartzlike structures.

15.51. Sieve is defined in Webster's New World Dictionary of the American Language, College Edition, The World Publishing Co., New York (1960) as "a utensil having many small meshed or perforated openings of a size allowing passage only to liquids or to the finer particles of loose or pulverized matter." Zeolites (with their various sized cavities that can trap or let pass a variety of molecules depending on their relative sizes) essentially do the same thing on a molecular level.

15.53. Gallium arsenide, GaAs, is isoelectronic with germanium and therefore would also serve as an intrinsic semiconductor. n-type semiconductors could be prepared from GaAs by doping it with atoms such as germanium or selenium that carry one more electron than gallium or arsenic, respectively. p-type semiconductors could similarly be prepared by doping GaAs with atoms such as zinc or germanium that carry one less electron than gallium or arsenic, respectively.

15.55. When sodium carbonate or limestone ($CaCO_3$) is heated (as they would be in the preparation of glass) they liberate carbon dioxide gas and sodium oxide and calcium oxide, respectively.

Chapter 16
Group 5A: The Pnicogens

The sections and subsections of this chapter are listed below.

16.1 Discovery and Isolation of the Elements
Antimony and Arsenic
Phosphorus
Bismuth
Nitrogen
16.2 Fundamental Properties and the Network
Uniqueness Principle
Other Network Components
Hydrides
Oxides and Hydroxides
Halides
16.3 A Survey of Nitrogen Oxidation States
Nitrogen (-3) Compounds: Nitrides and Ammonia
Nitrogen (-2), Hydrazine, N_2H_4
Nitrogen (-1), Hydroxylamine, NH_2OH
Nitrogen (+1), Nitrous Oxide, N_2O
Nitrogen (+2), Nitric Oxide, NO
Nitrogen (+3), Dinitrogen Trioxide, N_2O_3, and Nitrous Acid, HNO_2
Nitrogen (+4), Nitrogen Dioxide, NO_2
Nitrogen (+5), Dinitrogen Pentoxide, N_2O_5, and Nitric Acid, HNO_3
16.4 Reactions and Compounds of Practical Importance
Nitrogen Fixation
Nitrates and Nitrites
Matches and Phossy Jaw
Phosphates
16.5 Selected Topic in Depth: Photochemical Smog

Chapter Objectives

You should be able to

- briefly describe the relationship between some of the Group 5A elements and the ancient practice of alchemy
- briefly relate how and by whom the Group 5A elements were discovered
- outline the general manner in which the periodic law, uniqueness principle, inert pair effect, and the acid/base character of oxides are applicable to this group
- explain how the phosphazenes summarize the π bonding abilities of nitrogen and phosphorus
- compare and contrast the hydrides of the nitrogen, phosphorus, and arsenic
- represent the structural similarities among P_4, P_4O_6, and P_4O_{10}
- discuss and represent the structures of phosphoric acid, ortho-phosphate, and the various condensed phosphates
- discuss and represent the structures of phosphorous acid, ortho-phosphites and the various condensed phosphites
- discuss the structures, preparations, and properties of the phosphorus halides
- describe and draw the structure of borazine
- briefly describe the oxidation state, structure, properties, history, preparations, and reactions of
    1) ammonia
    2) hydrazine
    3) nitrous oxide
    4) nitric oxide
    5) nitrous acid and nitrites
    6) nitrogen dioxide
    7) nitric acid and nitrates
- briefly discuss three methods of nitrogen fixation
- explain why nitrogen compounds are often components of high explosives; give three examples
- briefly relate the history and hazards of matchmaking
- briefly explain the role of phosphates in fertilizers, food processing, soft drink formulations, and dental hygiene
- compare and contrast the general causes and characteristics of London and photochemical smog
- describe the daily production and abatement of both the primary and secondary pollutants that constitute photochemical smog

16.1. The name pnicogen means "choking producer." Obviously if we breathed in just nitrogen and not "common air" we would choke. If phosphorus is burned in air, it consumes the oxygen and makes the air a "choker." The same could be said of Priestley's test for the goodness of air, reacting it with nitric oxide. The result would be an "air" that would serve as a choker. Arsenic, the famous poison, and its compounds could also be loosely viewed as chokers as they cut off life.

16.3. $2Sb_2S_3(s) + 9O_2(g) \rightarrow 2Sb_2O_3(s) + 6SO_2(g)$
$2Sb_2O_3(s) + 3C(s) \rightarrow 4Sb(s) + 3CO_2(g)$

16.5.

$P_4(s)$	+	$5\,O_2(g)$	$\rightarrow P_4O_{10}(s)$
"cold fire"		"dephlogisticated air"	

16.7. The radii increase fairly steadily but not so much toward the bottom of the group. The ionization energies drop off quite regularly. The electronegativities and absolute values of the electron affinities also drop off as predicted but not so much at the bottom of the group. The irregularities in the radii, electronegativities, and electron affinities are due to the somewhat greater than expected effective nuclear charge experienced by the valence and incoming electrons. The filled nd and nf subshells do not shield these electrons as much as other subshells do.

16.9.

:N:::N:

sp - sp
N≡N
$p_x - p_x$
$p_y - p_y$

16.11. The unusually short P-O bond is due to the presence of a pπ-dπ bond between the unfilled d orbital of the phosphorus and a filled p orbital of the oxygen.

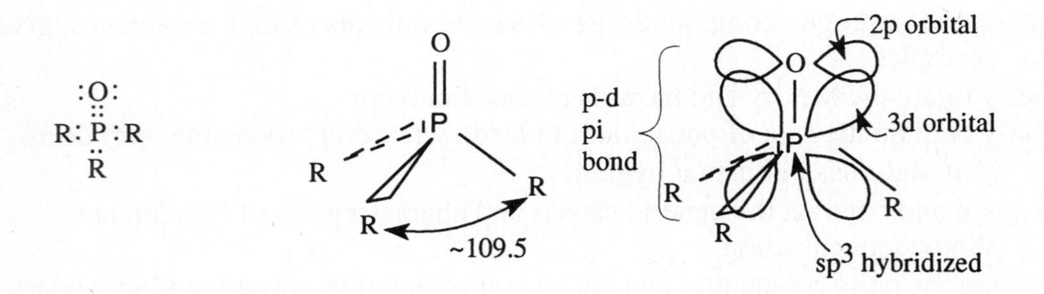

16.13. The Lewis and VSEPR structures of the cyclodiphosphazenes are shown in Figure 16.3(b). Note that the two equivalent resonance structures account for the nearly equal P-N distances in this cyclic molecule. The nitrogen atoms would be sp^2 hybridized and use their unhybridized 2p orbitals to form π bonds with adjacent phosphorus atoms. The phosphorus atoms, in turn, would be sp^3 hybridized and use their empty d orbitals to accept electron density from the nitrogens. The result would be delocalized pπ-dπ bonds around the 8-membered ring.

16.15. Of NH_3 and PH_3, phosphine is the stronger acid because the P-H bonds are longer and weaker than the N-H bonds of ammonia.

16.17. Phosphoric acid (H_3PO_4) versus arsenic acid (H_3AsO_4): Phosphorus is more electronegative than arsenic and therefore the P would draw more electron density from the O-H bond than would the As. The O-H bond of H_3PO_4 would be more susceptible to attack by water molecules and H_3PO_4 would be the stronger acid.

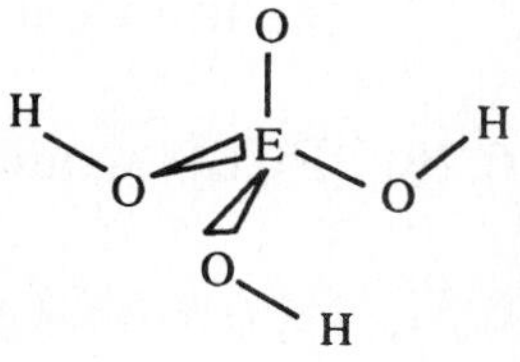

where E = As or P

16.19.
Although the question implies that either H_3PO_4-H_3PO_4 or H_3PO_4-H_2O hydrogen bonding occurs, it stands to reason that both exist in solutions of all percentages. It's just that H_3PO_4-H_2O hydrogen bonds predominate in solutions of less than 50%.

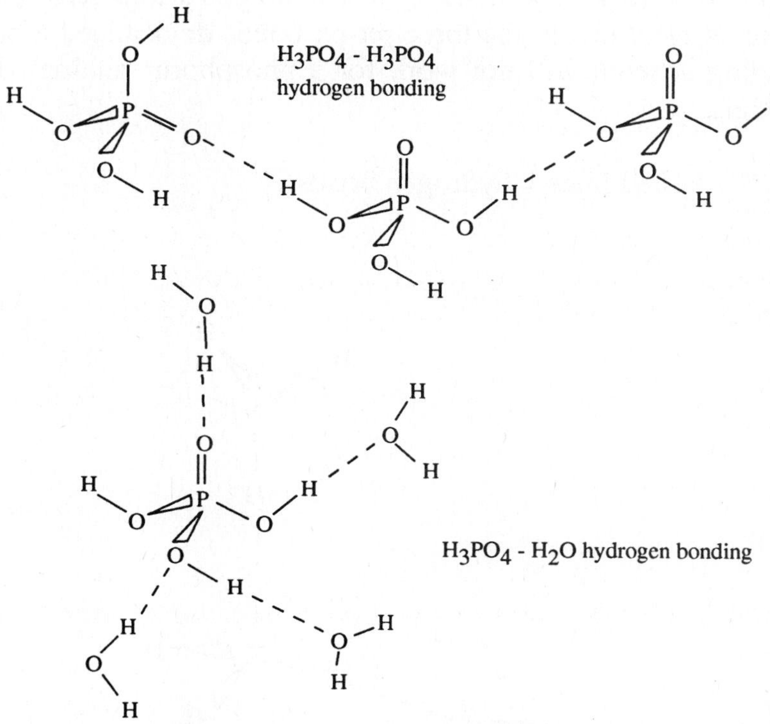

H_3PO_4 - H_3PO_4 hydrogen bonding

H_3PO_4 - H_2O hydrogen bonding

16.21. (a) potassium dihydrogenphosphate
(b) calcium dihydrogenphosphate
(c) magnesium pyrophosphate
(d) sodium trihydrogenpyrophosphite

16.23.

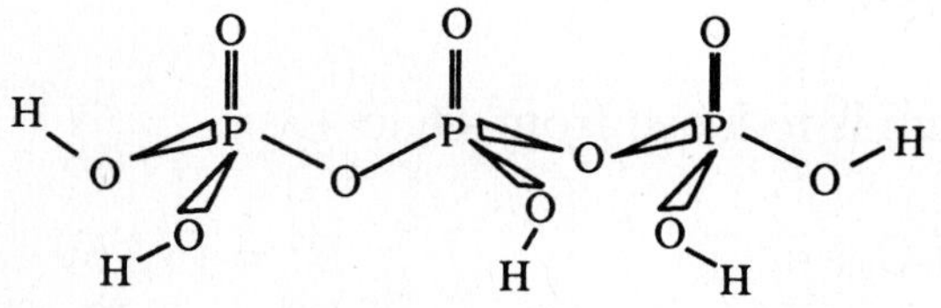

Tripolyphosphoric acid
Five acidic protons

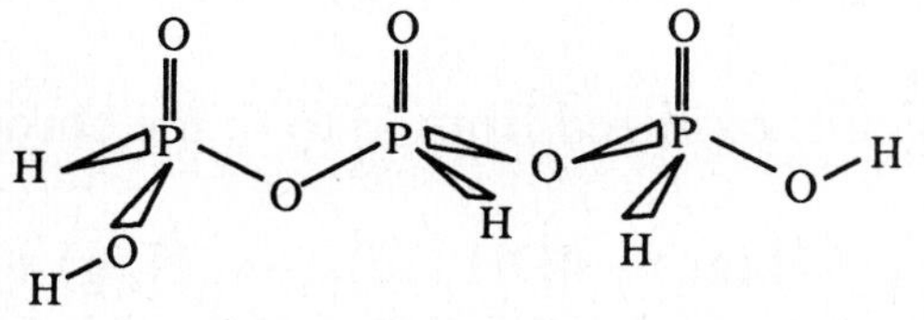

Tripolyphosphorous acid
Two acidic protons

16.25. NaH_2PO_2

16.27. (a) sodium dihydrogen arsenate
(b) silver(I) arsenite
(c) ammonium arsenate

16.29. $[PCl_4]^+$ would be tetrahedral while $[PCl_6]^-$ would be octahedral.

16.31. $P_4(s) + 6X_2(g) \rightarrow 4PX_3(g)$
(limiting reagent)

16.33. In borazine, both nitrogens and borons are sp^2 hybridized. The nitrogen atoms each donate a pair of electrons to the three pπ-pπ bonds delocalized around the 6-membered ring. A similar bonding scheme will not work for a phosphorus analog. Phosphorus is not capable of pπ-pπ bonding.

16.35. (dashed lines = hydrogen bonds)

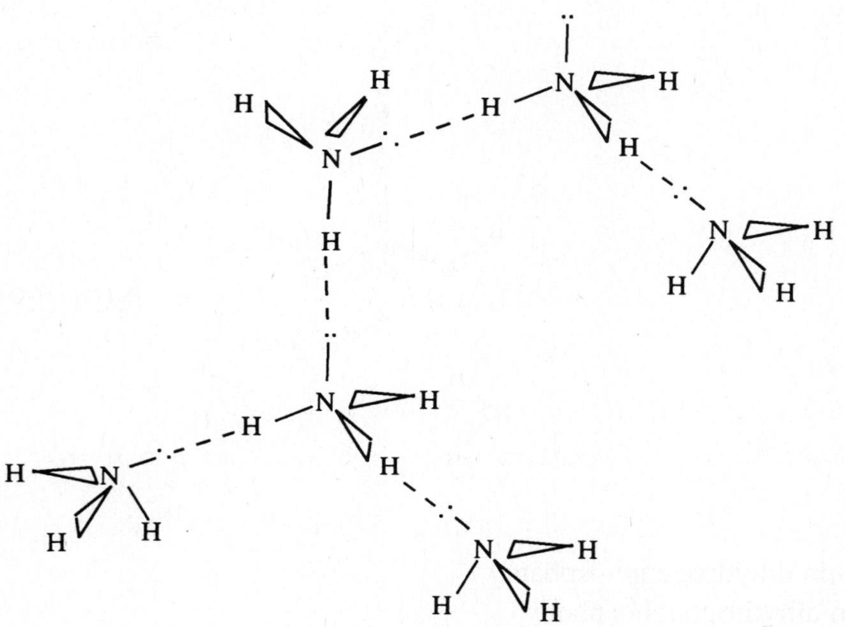

16.37.
$$\overset{-3\,+1}{(NH_4)_2}\overset{+6}{Cr_2}\overset{-2}{O_7}(s) \rightarrow \overset{0}{N_2}(g) + 4\overset{+1}{H_2}\overset{-2}{O}(g) + \overset{+3}{Cr_2}\overset{-2}{O_3}(s)$$

The nitrogen is oxidized from -3 to 0; the chromium is reduced from +6 to +3.

16.39.
$$3\,[\,N_2H_4(aq) + 4OH^-(aq) \rightarrow N_2(g) + 4H_2O + 4e^-\,] \qquad E° = +1.16v$$
$$4\,[\,MnO_4^-(aq) + 2H_2O(l) + 3e^- \rightarrow MnO_2(s) + 4OH^-(aq)\,] \qquad E° = +0.59v$$
$$3N_2H_4(aq) + 4MnO_4^-(aq) \rightarrow 3N_2(g) + 4H_2O(l) + 4MnO_2(s) + 4OH^-(aq) \qquad E° = +1.75v$$

Yes, hydrazine could accomplish this reduction under standard state conditions.

16.41. $\%O = \frac{16.00}{2(14.00) + 16.00} \times 100 = 36.4\%$

Air is only ~20% oxygen so there is a greater percentage in nitrous oxide. This calculation does support the observation that candles glow brighter in N_2O gas.

16.43.

H :Ö:N::N:Ö:H

Oxidation State of N = +2

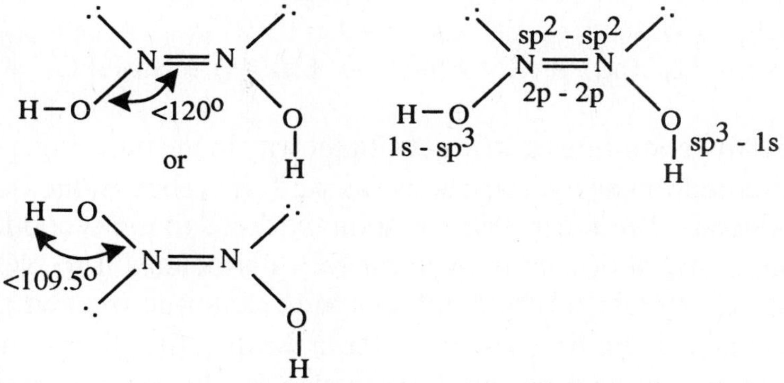

H−O, N=N, O−H; <120°

or

H−O, N=N, O−H; <109.5°

sp² - sp² (N=N), 2p - 2p; H−O 1s - sp³; O−H sp³ - 1s

16.45. Each NO molecule has an unpaired electron. When NO dimerizes to N_2O_2 these two unpaired electrons are used to establish an N-N bond. Bond formation is an exothermic process so the ΔH° is negative.

16.47.

:Ö::N:Ö:N::Ö:

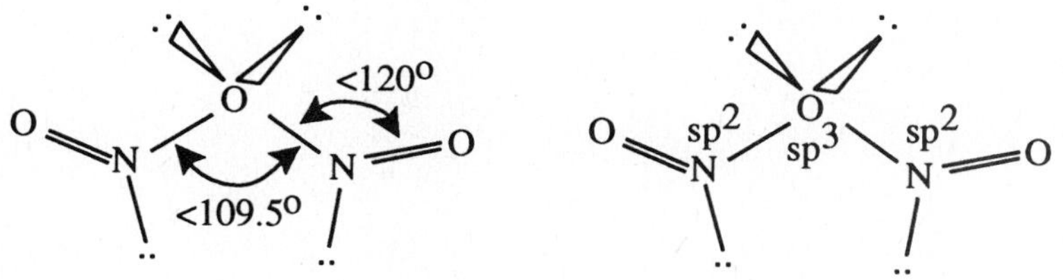

O=N−O−N=O; <120°; <109.5°

sp² (N), sp³ (O), sp² (N)

16.49.

$$1\,[NO_2 + H^+ + e^- \rightarrow HNO_2] \qquad E° = +1.10\text{v}$$
$$1\,[NO_2 + H_2O \rightarrow NO_3^- + 2H^+ + e^-] \qquad E° = -0.78\text{v}$$
$$2NO_2 + H_2O \rightarrow HNO_2 + NO_3^- + H^+ \qquad E° = +0.32\text{v}$$

This disproportionation reaction should occur under standard state conditions.

16.51.
$$S_8(s) + 48HNO_3(aq) \rightarrow 48NO_2(g) + 8SO_4^{2-}(aq) + 16H_3O^+(aq)$$
$$P_4(s) + 20HNO_3(aq) + 8H_2O \rightarrow 4PO_4^{3-}(aq) + 20NO_2(g) + 12H_3O^+(aq)$$

16.53.
$$Al(s) + 3HNO_3 + 3H_3O^+(aq) \rightarrow Al^{3+}(aq) + 3NO_2(g) + 6H_2O(l)$$
$$3Fe(s) + 2HNO_3 + 6H_3O^+(aq) \rightarrow 3Fe^{2+}(aq) + 2NO(g) + 10H_2O(l)$$

16.55.

$[\ddot{N}::C::\ddot{N}]^{2-}$

N=C=N, 180°

sp² - sp sp - sp²
N=C=N

16.57. (a) $3P_4(s) + 10ClO_3^- \xrightarrow{\Delta} 10Cl^- + 3P_4O_{10}$
(b) $3P_4S_3(s) + 16ClO_3^- \xrightarrow{\Delta} 16Cl^- + 3P_4O_{10} + 9SO_2$

16.59. $Ca(H_2PO_4)_2$ might be called monocalcium phosphate since it contains one calcium cation and phosphate. However this name might also be appropriate for $CaHPO_4$. A better name for the original compound would be calcium dihydrogenphosphate.

16.61. $CaHPO_4 \cdot 2H_2O(aq) + 2NaF(aq) \rightarrow CaF_2(s) + Na_2HPO_4 + 2H_2O$

16.63. As the daily commuter traffic gets under way in the morning, primary pollutants such as nitric oxide and unburned lower hydrocarbons (as well as carbon monoxide, not shown in Figure 16.12) would be produced. The nitric oxide is soon oxidized to the secondary pollutant nitrogen dioxide, NO_2. The photolytic action of the sun on NO_2 dissociates it to NO and the very reactive atomic oxygen. Atomic oxygen in turn combines with diatomic oxygen and the lower hydrocarbons to produce the oxidants including ozone. Later in the day, the afternoon traffic produces some primary pollutants again and some nitrogen dioxide also results. However, as the sun is setting late in the day, not many oxidants are produced the second time around.

Chapter 17
Sulfur, Selenium, Tellurium, and Polonium

The sections and subsections of this chapter are listed below.

17.1 Discovery and Isolation of the Elements
- Sulfur
- Tellurium and Selenium (Earth and Moon)
- Polonium

17.2 Fundamental Properties and the Network
- Hydrides
- Oxides and Oxoacids
- Halides

17.3 Allotropes and Compounds Involving Element-Element Bonds
- Allotropes
- Polycations and Anions
- Catenated Halides and Hydrides
- Catenated Oxoacids and Corresponding Salts

17.4 The Sulfur Nitrides

17.5 Reactions and Compounds of Practical Importance
- Sodium-Sulfide Batteries
- Photoelectric Uses of Selenium and Tellurium
- Sulfuric Acid

17.6 Selected Topic in Depth: Acid Rain

Chapter Objectives

You should be able to

- briefly relate how and by whom sulfur, selenium, tellurium, and polonium were discovered
- outline the general manner in which the periodic law, uniqueness principle, inert pair effect, acid-base character of oxides, and standard reduction potentials are applicable to the Group 6A elements
- summarize the preparations and properties of the noncatenated hydrides of sulfur and its heavier congeners
- summarize the preparations and properties of the oxides and noncatenated oxoacids of sulfur and its heavier congeners
- summarize the preparations and properties of the noncatenated halides of sulfur and its heavier congeners
- discuss the major allotropes of sulfur in the solid, liquid, and gaseous phases
- give some examples and discuss the properties and structures of some catenated polycations and anions of sulfur and its heavier congeners
- give some examples and discuss the properties, preparations, and structures of the catenated halides and hydrides of sulfur and its heavier congeners
- give some examples and discuss the properties, preparations, and structures of catenated sulfur-containing oxoacids and their corresponding salts
- give some examples and discuss the properties, preparations, and structures of some representative sulfur nitrides
- describe the nature of sodium-sulfide batteries
- describe the role of selenium and tellurium in the Xerox process and in II-VI semiconductors
- describe the industrial preparation and some of the uses of sulfuric acid
- describe the causes, consequences, and possible control of acid rain

17.1. Since gunpowder was introduced in the thirteenth century and matches soon after the discovery of phosphorus in late seventeenth century by Brandt, these inventors most likely did not regard sulfur as an element. Only in the first decade of the nineteenth century was sulfur so recognized.

17.3. The density of antimony is 6.69 g/cm^3 while that of tellurium is 6.24 g/cm^3. There is only a 7% difference between these two values so it is understandable that Müller confused them. Selenium, on the other hand, has a density of 4.79 g/cm^3, some 30% different from that of tellurium. A comparison of the densities of these two elements would easily distinguish them.

17.5. From sulfur on down the group, the chalcogens have less and less affinity for electrons. This trend is demonstrated in the values becoming less negative, sulfur to tellurium and reflects the fact that the incoming electron is farther and farther away from the effective nuclear charge. Another factor, perhaps reflected in the small drop off of the values toward the bottom of the group, is the slightly increased values of Z_{eff} after an nd^{10} or nf^{14} subshell is filled. The magnitude of the electron affinity of oxygen is smaller than expected due to the high degree of electron-electron repulsion in the very small oxygen atom.

17.7. Sulfur is a more representative chalcogen than oxygen (re: uniqueness principle). Oxygen is considerably smaller than sulfur and therefore is capable of $p\pi$-$p\pi$ bonding. This makes diatomic oxygen gas the standard state for this element whereas catenated chains and rings (composed of single bonds) are more typical for sulfur and most of the other chalcogens (except polonium). The small size of oxygen also results in hindered rotation about O-O single bonds not typical of E-E bonds in sulfur or the heavier chalcogens. The high electronegativity of oxygen results in a hydride, H_2O, which is unique compared to the hydrides of the other chalcogens. Sulfur is also more typical of selenium and tellurium in its ability to form (1) a variety of oxoacids characterized by both E-O-E and E-E bonds and (2) EO_2 and EO_3 oxides that are acid anhydrides. These and other compounds of sulfur and its heavier congeners are characterized by $p\pi$-$d\pi$ bonds, something that oxygen is incapable of due to the relative unavailability of its d orbitals.

17.9. H_2S should be a better acid than water because the S-H bonds are longer and weaker than the O-H bonds of water. The latter factor is more important than the decrease in $EN_{chalcogen}$ and the corresponding decrease in the polarity of the hydrogen-chalcogen bond which would make it less susceptible to attack by polar water molecules.
Data: K_w of $H_2O = K_{a1} = 1.0 \times 10^{-14}$; K_{a1} of $H_2S = 1.0 \times 10^{-7}$

17.11. $Al_2Se_3(s) + H_2O(l) \rightarrow H_2Se(g) + Al_2O_3(s)$
$FeS(s) + HCl(aq) \rightarrow H_2Se(g) + FeCl_2(aq)$

17.13. $FeS_2(s) + 2O_2(g) \rightarrow 2SO_2(g) + Fe(s)$
$HgS(s) + O_2(g) \rightarrow SO_2(g) + Hg(l)$

17.15. $SO_2{\cdot}H_2O(aq) + H_2O \rightleftharpoons HSO_3^-(aq) + H_3O^+(aq)$

17.17. $SO_3^{2-}(aq) + H_2O \rightleftharpoons HSO_3^-(aq) + OH(aq)$

$$K_b = \frac{K_w}{K_{a2}} = \frac{[H^+][OH^-]}{\frac{[H^+][SO_3^{2-}]}{[HSO_3^-]}} = \frac{[HSO_3^-][OH^-]}{[SO_3^{2-}]} = \frac{1.00 \times 10^{-14}}{6.3 \times 10^{-8}} = 1.6 \times 10^{-7}$$

17.19. $SO_3^{2-}(aq) + Cl_2(g) \rightarrow SO_4^{2-}(aq) + Cl^-(aq)$ (unbalanced)

Balance by half equations:

$$SO_3^{2-} + H_2O \rightarrow SO_4^{2-} + 2H^+ + 2e^- \quad 1$$
$$Cl_2 + 2e^- \rightarrow 2Cl^- \quad 1$$
$$SO_3^{2-} + Cl_2 + H_2O \rightarrow SO_4^{2-} + 2H^+ + 2Cl^-$$

Given that H^+ is a product, high acid concentrations would shift the position of equilibrium of this reaction back toward the left. This is an example of Le Châtelier's Principle.

17.21. Sulfuric acid has one more nonhydroxyl-group oxygen atom withdrawing electron density from the O-H bonds. This makes these bonds more polar and therefore more susceptible to attack by polar water molecules. It follows that H_2SO_4 is the stronger acid.

17.23. $2e^- + S(s) + 2H^+(aq) \rightarrow H_2S(g)$ $E° = 0.14v$; $\Delta G° = -(2)(96.5kJ/v)(0.14v) = -27kJ$

$SO_4^{2-}(aq) + 8H^+(aq) + 6e^- \rightarrow S(s) + 4H_2O(l)$ $E° = 0.37v$; $\Delta G° = -(6)(96.5kJ/v)(0.37v) = -214\ kJ$

$8e^- + 10H^+(aq) + SO_4^{2-}(aq) \rightarrow H_2S(g) + 4H_2O(l)$ $\Delta G° = -241\ kJ$

$$E° = -\frac{-241kJ}{(8)(96.5kJ/v)} = 0.31v$$

17.25. (a)

$$Sn(s) \rightarrow Sn^{2+}(aq) + 2e^- \qquad E° = +0.14v$$
$$2e^- + 2H^+(aq) \rightarrow H_2(g) \qquad E° = 0.00v$$
$$SO_4^{2-}(aq) \rightarrow SO_4^{2-}(aq)$$
$$Sn(s) + 2H^+(aq) + SO_4^{2-}(aq) \rightarrow Sn^{2+}(aq) + H_2(g) + SO_4^{2-}(aq) \qquad E° = +0.14v$$

(b)

$$Sn(s) \rightarrow Sn^{2+}(aq) + 2e^- \qquad E° = +0.14v$$
$$Sn^{2+}(aq) \rightarrow Sn^{4+}(aq) + 2e^- \qquad E° = -0.13v$$
$$2\,[SO_4^{2-}(aq) + 4H^+(aq) + 2e^- \rightarrow SO_2(g) + 4H_2O(l)\,] \qquad E° = +0.20v$$
$$Sn(s) + 2SO_4^{2-}(aq) + 8H^+(aq) \rightarrow 2SO_2(g) + 4H_2O(l) + Sn^{4+}(aq) \qquad E° = +0.21v$$

17.27. Fluorine is isoelectronic with OH so fluorosulfuric acid is analogous with sulfuric acid but with an OH replaced by an F. Fluorine is more electronegative than oxygen so even more electron density is removed from the one remaining O-H bond than in sulfuric acid. That bond is therefore more polar and more susceptible to attack by polar water molecules. (The Lewis structure could also be written with S=O bonds with the sulfur having an expanded octet.)

:F:
:O:S:O:
:O:
H

17.29. Sulfamic acid, $(H_2N)SO_3H$ (a)

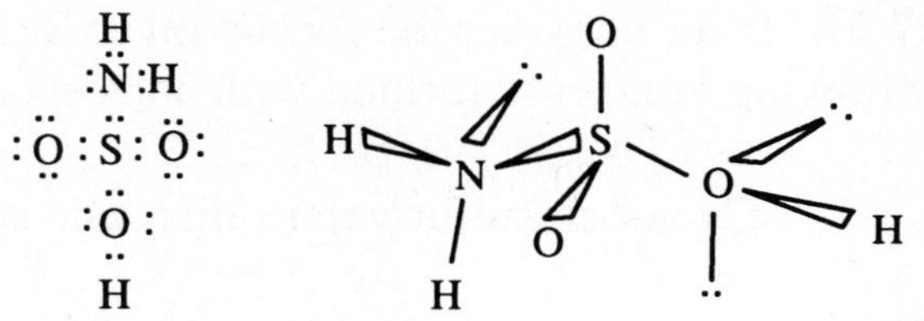

(The Lewis structure could also be written with S=O bonds with the sulfur having an expanded octet.)

(b) The H of the O-H is acidic while the $-NH_2$ group is basic, therefore a proton can be transferred from the O-H to the NH_2.

17.31.

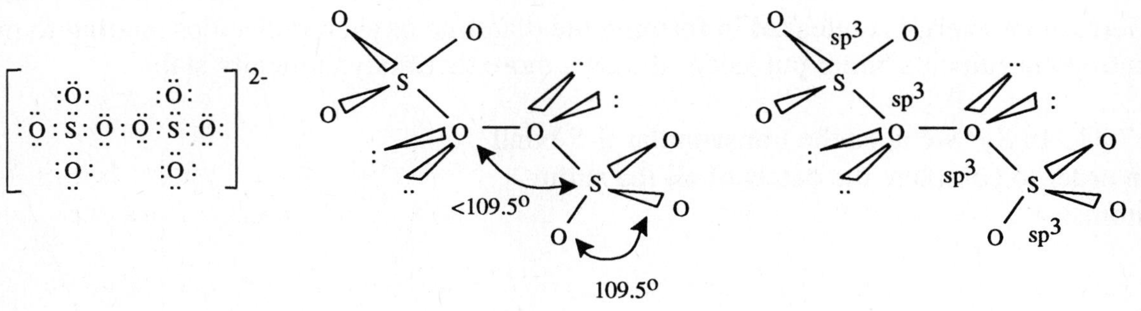

<109.5°

109.5°

In the VBT diagram all the bonds can be pictured as being between sp^3 hybrid orbitals although some chemists may prefer to picture the terminal S-O bonds as being due to sp^3-p overlap.

17.33.

$2NaHSO_3(s) \xrightarrow{\Delta} Na_2S_2O_5(s) + H_2O(l)$

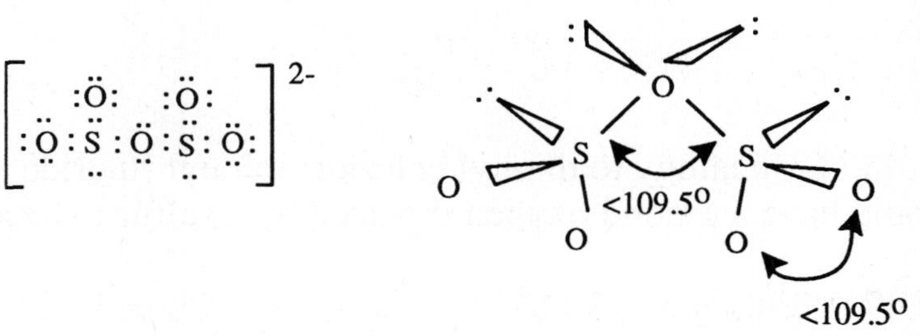

<109.5°

<109.5°

17.35. The lone pair, being confined by only one nucleus, will spread out and occupy more volume than a bonding pair (confined by two nuclei). In structure (a), there are only two F's at 90° to the lone pair whereas in structure (b) there are three. Therefore, there is more room for the lone pair to spread out in structure (a).

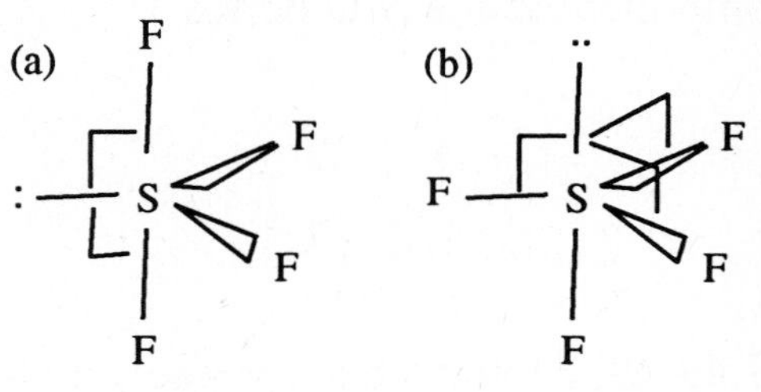

(a) (b)

17.37. The Po^{IV} ion has a higher charge density. Therefore, we have come to expect that its compounds will have a greater covalent character than those of Po^{II}.

17.39. If we consider the formation of S_8 and 4 O_2's from the respective free atoms, the following bonds are formed with corresponding amounts of energy released:

Cyclooctasulfur versus diatomic sulfur:	8 S-S	8 x 226 = 1808 kJ
	4 S=S	4 x 423 = 1692 kJ

Therefore more energy is released in forming cyclooctasulfur and this would tend to make (entropy arguments aside) the S_8 more thermodynamically stable.

"Cyclooctaoxygen" versus diatomic oxygen:	8 O-O	8 x 146 = 1168 kJ
	4 O=O	4 x 496 = 1984 kJ

Here, more energy is released in forming the diatomic oxygen molecules tending to make (again entropy arguments being put aside) the O_2's more thermodynamically stable.

17.41. In S_8^{2+} we need the transannular S-S bond in order to complete the octets of all the sulfur atoms.

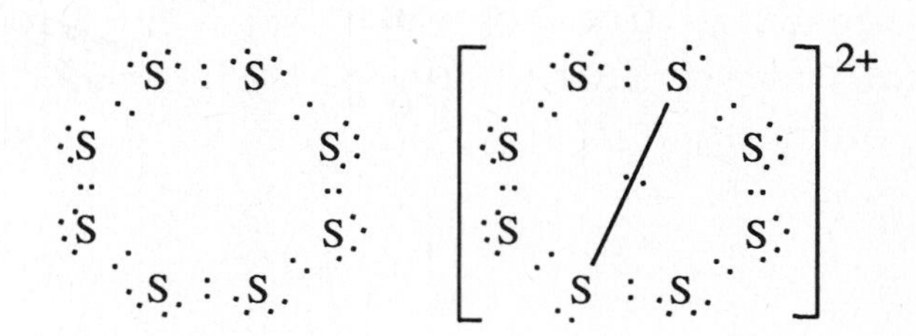

17.43.
There would most likely be fairly free rotation about the longer (compared to the O-O bond in H_2O_2, for example) S-S bond.

17.45. By analogy to thionyl chloride, thionyl fluoride would be SOF_2. Thiothionyl fluoride would have a second oxygen replaced by a sulfur to become SSF_2.

17.47. Trithionate, $S_3O_6^{2-}$

(There may also be some pπ-dπ component to the S-O bonds.)

17.49. S_8 contains 8 x 6 = 48e$^-$ and therefore has two more electron pairs than S_4N_4. This would result in two less transannular S-S bonds in the S_8 molecule as compared to S_4N_4. S_8^{2+} contains (8 x 6) - 2 = 46e$^-$, one more electron pair than S_4N_4, resulting in one less transannular S-S bond in the S_8^{2+} cation as compared to S_4N_4. Resonance structures for S_8^{2+} and S_8 are shown below.

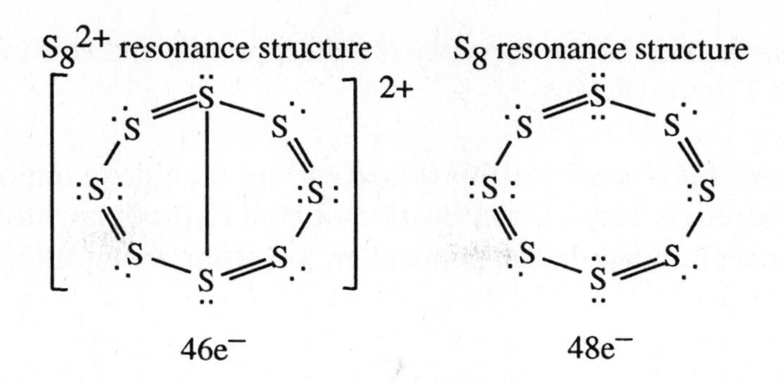

17.51. In S_2N_2, there are (6x2) + (5x2) = 22e- to be accounted for. The four resonance structures drawn at the right show how those 22 electrons might be distributed in the molecule. Each S-N bond would have a bond order of 1 1/2.

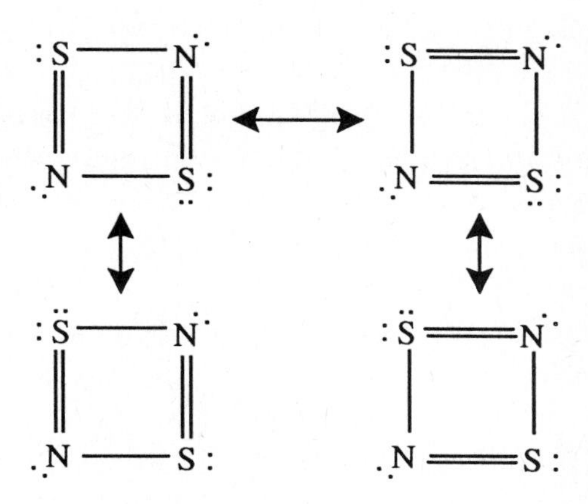

17.53. Germanium is an intrinsic semiconductor of the type covered in Chapter 15 (p. 401). GaAs is a III-V intrinsic semiconductor isoelectronic with Ge (or 2Ge). ZnSe is a II-VI intrinsic semiconductor also isoelectronic with germanium.

17.55. $2SO_2(g) + O_2(g) \rightarrow 2SO_3(g)$

From a kinetic point of view high temperatures would increase the rate of this reaction. High temperatures increase the number and impact of each molecular collision resulting in more collisions that are able to move the reaction over its energy of activation (E_a). From a thermodynamic point of view a high pressure would favor the above reaction going left to right as written. Reactions at equilibrium tend to proceed in a direction that offset a stress (LeChâtelier's principle). If the stress is high pressure this reaction will go left to right toward the side with fewer gaseous substances. The optimum temperature thermodynamically depends on whether a given reaction is exo- or endothermic. This reaction is endothermic and therefore will be shifted to the right at high temperatures. A high $[O_2]$ will also favor the left to right reaction again due to LeChâtelier's principle.

17.57. $SO_3(g) + D_2O(l) \rightarrow D_2SO_3(aq)$

17.59. Acid rain is a greater problem in the northeastern parts of the United States because (a) prevailing winds sweep the $SO_2/SO_3/H_2SO_4$ (as well as NO_x / HNO_3) from stationary sources (including many electrical-generating power plants that burn relatively high-sulfur coal) in the midwest toward the northeast and (b) the capacity of lakes, ponds, and soil in the northeast to neutralize acid is low (see Figure 17.10).

17.61. The burning of fossil fuels would still produce carbon dioxide, a major greenhouse gas. If we were to continue to strongly rely on combustion of coal to produce our energy, enormous amounts of this gas might accumulate in the atmosphere resulting in significant global warming.

Chapter 18
Group 7A: The Halogens

The sections and subsections of this chapter are listed below.

18.1 Discovery and Isolation of the Elements
- Chlorine
- Iodine
- Bromine
- Fluorine
- Astatine

18.2 Fundamental Properties and the Network
- Hydrides
- Halides
- Oxides

18.3 Oxoacids and Their Salts
- Hypohalous Acids, HOX, and Hypohalites, OX^-
- Halous acids, HOXO and Halites, XO_2^-
- Halic acids, $HOXO_2$ and Halates, XO_3^-
- Perhalic acids, $HOXO_3$ and Perhalates, XO_4^-

18.4 Neutral and Ionic Interhalogens

18.5 Reactions and Compounds of Practical Importance
- Fluoridation
- Chlorination
- Bleaches
- Bromides

18.6 Selected Topic in Depth: Chlorofluorocarbons (CFCs): A Threat to the Ozone Layer

Chapter Objectives

You should be able to

- briefly relate how and by whom the halogens were discovered
- outline the general manner in which the periodic law, uniqueness principle, acid-base character of oxides, and standard reduction potentials are applicable to the group
- explain why fluorine is such an extraordinarily powerful oxidizing agent and how this property made it so difficult to isolate
- discuss the preparations and properties of the hydrogen halides
- summarize and give representative examples of the various ways to prepare halides of the representative elements
- list and discuss the major applications of the halogen oxides
- discuss the preparations, properties, and applications of
  1) the hypohalous acids and the hypohalites
  2) the halous acids and the halites
  3) the halic acids and the halates
  4) the perhalic acids and the perhalates
- discuss the preparation, structures, and major reactions of the interhalogens
- briefly explain the role of the halogens in the fluoridation and chlorination of water supplies, the nature of bleaches, and the applications of the bromides
- chemically explain the threat that the chlorofluorocarbons pose to the ozone layer

18.1. $NaCl + HX \rightarrow HCl(g) + NaX$

strong acid (HX)

18.3. Arrhenius acids are those that liberate $H^+(aq)$ ions in solution so Arrhenius would agree that hydrogen is a necessary component of an acid. Brønsted-Lowry acids are proton (H^+) donors so they also would see hydrogen content as characteristic of an acid. Lewis acids, on the other hand, are electron pair acceptors. Compounds need not contain hydrogen to be Lewis acids so Lewis would disagree with Davy.

18.5. Since thyroxine is an iodine-containing amino acid produced in the thyroid gland, radioactive iodine 131 goes directly to the thyroid and can be used to treat cancer in that gland.

$$^{131}_{53}I \rightarrow {}^{0}_{-1}e + {}^{131}_{54}Xe$$

18.7. ICl MW = 126.9 + 35.45 = 162.4u

Br_2 MW = 2(79.90) = 159.8u

They differ by only 1 or 2 percent.

18.9. $^{211}_{85}At \xrightarrow{\alpha} {}^{4}_{2}He + {}^{207}_{83}Bi$

18.11. (a) $Cl_2O_7 + H_2O \rightarrow 2HClO_4$, perchloric acid

(b) $Br_2O + H_2O \rightarrow 2HBrO$, hypobromous acid

(c) $I_2O_3 + H_2O \rightarrow 2HIO_2$, iodous acid

18.13. HF is a weak acid because the H-F bond is considerably shorter and stronger than the other H-X bonds (re: uniqueness principle). Therefore, more energy is required to break the H-F bond and HF is a weaker acid.

18.15. $H\text{-}H(g) + F\text{-}F(g) \xrightarrow{\Delta H} 2H\text{-}F(g)$

Bonds broken	Bonds formed
1H-H = 436 kJ/mol	2H-F = 2(568)
1F-F = 151 kJ/mol	= 1136 kJ/mol
587 kJ/mol	

$\Delta H = +587 - 1136 = -549$ kJ/mol

$\Delta H^\circ_f = 1/2\Delta H = -274$ kJ/mol

18.17. Oxygen has a lower than (that is, not as negative as) expected electron affinity because it is so small. An electron added to this neutral atom experiences a significant amount of electron-electron repulsion due to the other six electrons already in place in the small volume characteristic of the 2s and 2p orbitals. The positive contribution to the potential energy by this electron-electron repulsion makes it more difficult to add an electron to the oxygen atom than would be expected on the basis of factors such as effective nuclear charge.

18.19.

$$
\begin{array}{ccc}
1/2\ F_2(g) + e^- & \xrightarrow{\Delta H^\circ_{red}} & F^-(aq) \\
\Delta H(g) \downarrow & & \uparrow \Delta H_{hyd} \\
F(g) & \xrightarrow{EA} & F^-(g)
\end{array}
$$

$$\Delta H^\circ_{red} = \Delta H(g) + EA + \Delta H_{hyd}$$

The above reaction is very thermodynamically spontaneous because (a) $\Delta H(g)$, corresponding to breaking of the weak F-F bond of F_2, is so low and (b) the ΔH_{hyd} of $F^-(g)$ is so large and negative. The latter is due to the very high charge density of F^- leading to a very strong and exothermic interaction with the polar water solvent.

18.21. $NaF(aq) + H_2SO_4(aq) \rightarrow 2HF(aq) + Na_2SO_4(aq)$
$NaCl(aq) + H_2SO_4(aq) \rightarrow 2HCl(aq) + Na_2SO_4(aq)$
$2NaBr(aq) + 2H_2SO_4(aq) \rightarrow Br_2(l) + SO_2(g) + 2H_2O(l) + Na_2SO_4(aq)$
$2NaI(aq) + 2H_2SO_4(aq) \rightarrow I_2(s) + SO_2(g) + 2H_2O(l) + Na_2SO_4(aq)$

Sulfuric acid is a strong enough oxidizing agent to convert bromides to bromine and iodides to iodine but it is not strong enough to produce fluorine and chlorine from the fluorides and chlorides, respectively.

18.23.

$2[2I^-(aq) \rightarrow I_2(s) + 2e^-]$	$E^\circ = -0.54v$
$\underline{1[O_2(g) + 4H^+(aq) + 4e^- \rightarrow 2H_2O(l)]}$	$\underline{E^\circ = +1.23v}$
$4HI(aq) + O_2(g) \rightarrow 2I_2(s) + 2H_2O(l)$	$E^\circ = +0.69v$

$2[2Br^-(aq) \rightarrow Br_2(l) + 2e^-]$	$E^\circ = -1.07v$
$\underline{1[O_2(g) + 4H^+(aq) + 4e^- \rightarrow 2H_2O(l)]}$	$\underline{E^\circ = +1.23v}$
$4HBr(aq) + O_2(g) \rightarrow 2Br_2(l) + 2H_2O(l)$	$E^\circ = +0.16v$

Both of these reactions are spontaneous under standard state conditions. The hydrohalic acids are not immediately converted to the halogens because the concentration or partial pressure of $O_2(g)$ is not great enough in aqueous solutions.

18.25. Note in the sketch of the $(HF)_6$ hexamer shown at the right that the H-F—H hydrogen bonds must be linear.

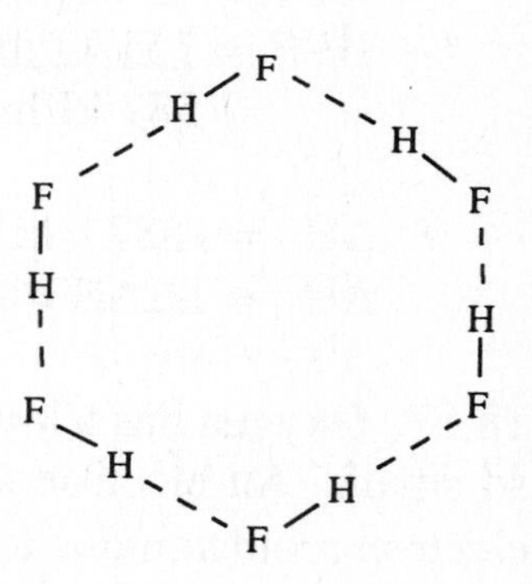

18.27. $2\,[NO_3^-(aq) + 4H^+(aq) + 3e^- \rightarrow NO(g) + 2H_2O]$ $\qquad E° = 0.96v$

$\underline{\quad 2\,[\,2Cl^-(aq) \rightarrow Cl_2(g) + 2e^-\,]}$ $\qquad \underline{E° = -2.87v}$

$2NO_3^-(aq) + 8H^+(aq) + 6Cl^-(aq) \rightarrow 2NO(g) + 3Cl_2(g) + 4H_2O$ $\qquad E° = -1.91v$

No, HNO_3 could not be used to produce chlorine from chloride.

18.29. $Al_2O_3(s) + 6HF(g) \rightarrow 2AlF_3(s) + 3H_2O(l)$
$2Al(s) + 3Cl_2(g) \rightarrow 2AlCl_3(s)$

18.31. (a) $2Co_3O_4(s) + 6ClF_3(g) \rightarrow 6CoF_3(s) + 3Cl_2(g) + 4O_2(g)$
(b) $2B_2O_3(s) + 4BrF_3(l) \rightarrow 4BF_3(g) + 2Br_2(l) + 3O_2(g)$
(c) $3SiO_2(s) + 4BrF_3(l) \rightarrow 3SiF_4(g) + 2Br_2(l) + 3O_2(g)$

18.33. (a) $GeCl_2(s) + 2H_2O \rightarrow Ge(OH)_2(aq) + 2HCl(aq)$
(b) $PCl_3(s) + 3H_2O \rightarrow H_3PO_3(aq) + 3HCl(aq)$
(c) $PCl_5(s) + 4H_2O \rightarrow H_3PO_4(aq) + 5HCl(aq)$

18.35. $NaSCN(aq) + AgNO_3(aq) \rightarrow AgSCN(s) + NaNO_3(aq)$

18.37. Cl_2O $\qquad ClO_2$

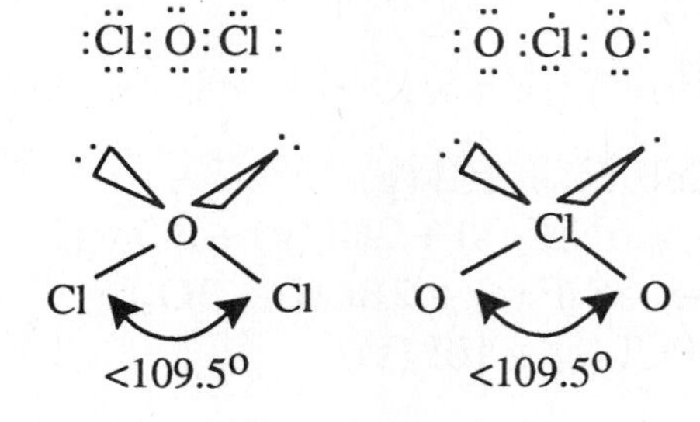

18.39. Oxidizing agent: I_2O_5
Reducing agent: CO

18.41. Equation (18.30) is a redox reaction with X_2 being both the oxidizing and the reducing agent. That is, X_2 disproportionates to HOX and X^-.

18.43. $Ba(BrO_3)_2(aq) + H_2SO_4(aq) \rightarrow BaSO_4(s) + 2HBrO_3(aq)$
This is a so-called "milkshake" reaction.

18.45. $HIO_3 + 5HI \rightarrow 3I_2 + 3H_2O$
Oxidizing agent: HIO_3
Reducing agent: HI

18.47. ${}^{83}_{34}SeO_4^{2-} \xrightarrow{\beta^-} {}^{83}_{35}BrO_4^- + {}^{0}_{-1}e$

18.49. $1\,[\,BrO_3^-(aq) + H_2O(l) \rightarrow BrO_4^-(aq) + 2H^+(aq) + 2e^-\,]$ $E° = -1.74v$

$1\,[XeF_2(aq) + 2H^+(aq) + 2e^- \rightarrow Xe(g) + 2HF(aq)\,]$ $E° = +2.64v$

$XeF_2(aq) + H_2O(l)(aq) + BrO_3^-(aq) \rightarrow Xe(g) + 2HF(aq) + BrO_4^-(aq)$ $E° = +0.90v$

The above reaction is thermodynamically feasible.

18.51. $H_4I_2O_9$

This is a condensation dimer of HIO_4 and H_5IO_6.

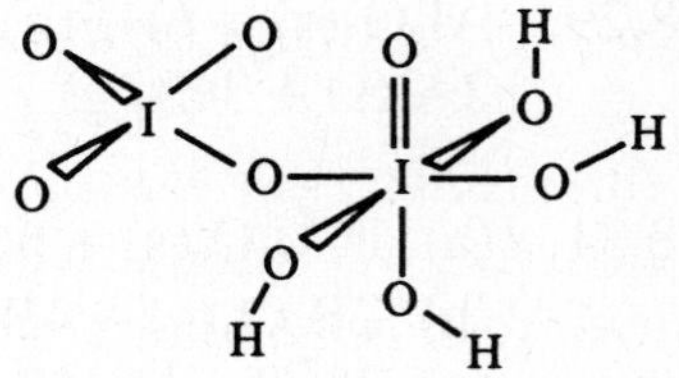

$H_7I_3O_{14}$

This is a condensation trimer of HIO_4 and two H_5IO_6's.

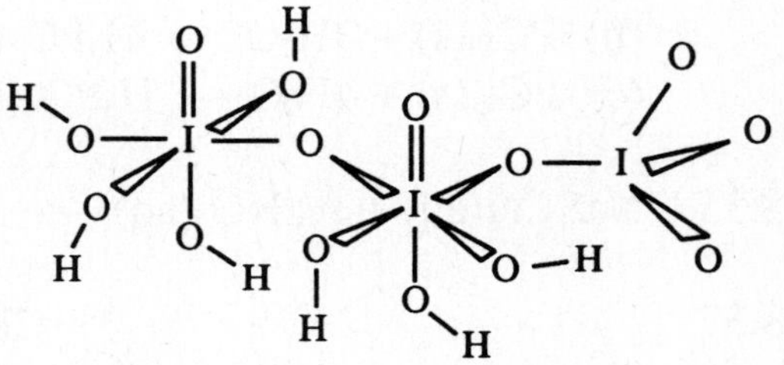

18.53. $Br_2(s) + F_2(g) \rightarrow 2BrF(g)$

$Br_2(s) + 3F_2(g) \rightarrow 2BrF_3(l)$

$BrF_3(l) + F_2(g) \rightarrow BrF_5(l)$

18.55. (a) $Se(s) + 4ClF(g) \rightarrow SeF_4(s) + 2Cl_2(g)$

(b) $6NiO(s) + 4ClF_3(g) \rightarrow 6NiF_2(s) + 2Cl_2(g) + 3O_2(g)$

(c) $3SiO_2(s) + 4BrF_3(l) \rightarrow 3SiF_4(s) + 2Br_2(l) + 3O_2(g)$

(d) $P_4(s) + 20ICl(s) \rightarrow 4PCl_5(s) + 10I_2(s)$

18.57. $IF_3 + 2H_2O \rightarrow 3HF + HIO_2$

$IF_5 + 3H_2O \rightarrow 5HF + HIO_3$

$IF_7 + 4H_2O \rightarrow 7HF + HIO_4$

18.59

(a)

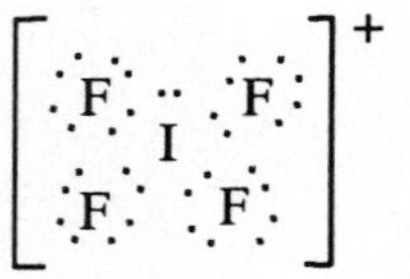

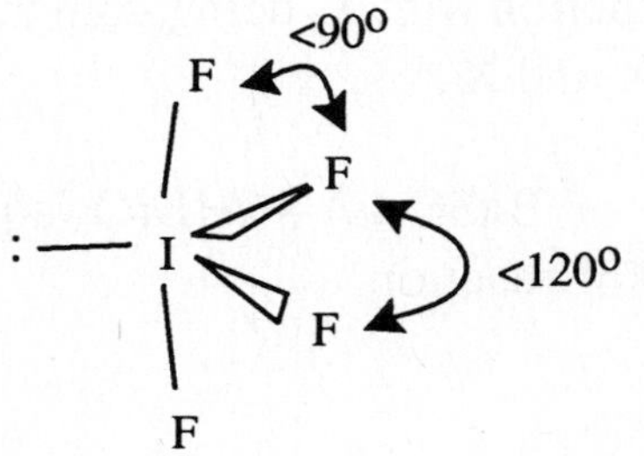

(b)

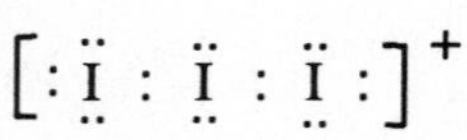

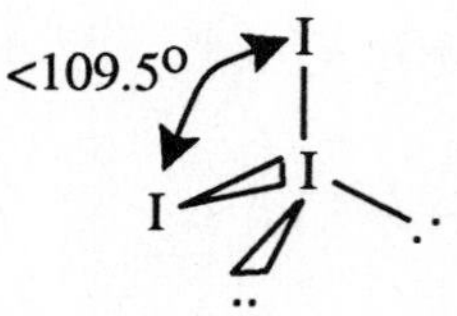

18.61. Fluorine 20 can be incorporated in small amounts into teeth and bones in the form of fluoroapatite, $Ca_{10}(PO_4)_6{}^{20}F_2$. For a given geographical area with a given amount of fluoride in the water supply, a hominid should have a predictable amount of fluoride in his or her bones or teeth. Knowing the isotopic abundance of fluorine 20, one could calculate the expected levels of F 20 in the teeth or bones at death. After death, the amount of F 20 would decrease. Measurement of the existing level of F 20 (by counting beta minus particles), combined with knowledge of the half-life of F 20 would yield a determination of the age of the sample.

18.63. Hydrogen peroxide is a good bleach because it is a strong oxidizing agent. H_2O_2 takes electrons away from organic dyes (responsible for many colors) such that transitions among energy levels resulting in absorption of visible frequencies are no longer possible.

$$\underset{\text{electrons taken from dyes}}{2e^-} + 2H^+ + H_2O_2 \rightarrow 2H_2O$$

18.65. $CFCl_3 \xrightarrow{h\nu} Cl + CFCl_2$

18.67. The Antarctic gets colder than the Arctic and the Antarctic polar vortex is more stable. This means that during the Antarctic winter, inert chlorine reservoirs can be readily set up from which active ozone-destroying chlorine and chlorine compounds can be released in the polar spring.

Chapter 19
Group 8A: The Noble Gases

The sections and subsections of this chapter are listed below.

19.1 Discovery and Isolation of the Elements
- Argon
- Helium
- Krypton, Neon, and Xenon
- Radon

19.2 Fundamental Properties and the Network

19.3 Compounds of Noble Gases
- History
- Fluorides
- Structures
- Other Compounds

19.4 Physical Properties and Elements of Practical Importance

19.5 Selected Topic in Depth: Radon as a Carcinogen

Chapter Objectives

You should be able to

- briefly relate how and by whom the noble gases were discovered
- outline the general manner in which the periodic law is applicable to this group
- briefly relate how Bartlett came to prepare the first compound of xenon
- discuss the preparations, properties, and structures of some representative fluorides and oxides of xenon
- give several examples of compounds of the noble gases other than xenon
- list a few representative uses of the noble gases
- relate and discuss the discovery and mechanism of radon as a major environmental health problem

19.1. The noble metals, like the noble gases, are generally reluctant to react with any other element.

19.3. $Mg(ClO_4)_2 + 6H_2O(g) \rightarrow Mg(ClO_4)_2 \cdot 6H_2O$ (see p. 304 for further details)

19.5. $\dfrac{\text{AW of oxygen}}{\text{AW of hydrogen}} = \dfrac{15.9994u}{1.00797u} = 15.8729$ This value is within 1% of the integer 16.

19.7. $3Mg(s) + N_2(g) \rightarrow Mg_3N_2(s)$ (see p. 303 for further details)

19.9. $\left(\dfrac{3.70g\ Kr}{L}\right)\left(\dfrac{22.4L}{mol}\right) = 82.9$ g/mol

19.11. The electron affinities increase in value (that is, there is a lesser tendency to add an electron) going down the noble gases because the electron being added to each atom must occupy a position farther and farther away from an approximately constant effective nuclear charge.

19.13. The effective nuclear charge felt by the valence electrons of a noble gas atom is very large. Therefore, these electrons are very difficult to remove. The Z_{eff} felt by an incoming electron (that would be added to an ns orbital outside the closed valence shell of electrons) is, on the other hand, very small. Accordingly, it is difficult to add an electron to these atoms. So we see that the $2s^22p^6$ configuration is particularly stable because it is difficult to either add to or remove electrons from it.

19.15. $UO_3(s) + 6HF(g) \rightarrow UF_6(l) + 3H_2O(l)$ or
$2UO_3(s) + 4ClF_3(g) \rightarrow 2UF_6(l) + 2Cl_2(g) + 3O_2(g)$

19.17. $\overset{0}{O_2}(g) + \overset{+6\ -1}{PtF_6}(g) \rightarrow \overset{+1/2\ +5\ -1}{O^+_2PtF_6}$

oxygen 0 → +1/2 oxidized, reducing agent
Pt +6 → +5 reduced, oxidizing agent

19.19. $Xe + RuF_6 \rightarrow Xe^+RuF_6^-$
$Xe + RhF_6 \rightarrow Xe^+RhF_6^-$

19.21. $4KI(s) + XeF_4(s) \rightarrow 4KF(s) + Xe(g) + 2I_2(s)$

19.23. Oxidizing agent: XeF_2
Reducing agent: H_2O

19.25. $2SbF_5(s) + XeF_2 \rightarrow XeF^+ Sb_2F_{11}^-$
$3Xe + XeF^+Sb_2F_{11}^- + 2SbF_5 \rightarrow 2Xe_2^+Sb_2F_{11}^-$

19.27. These structures are all consistent with the VSEPRT.

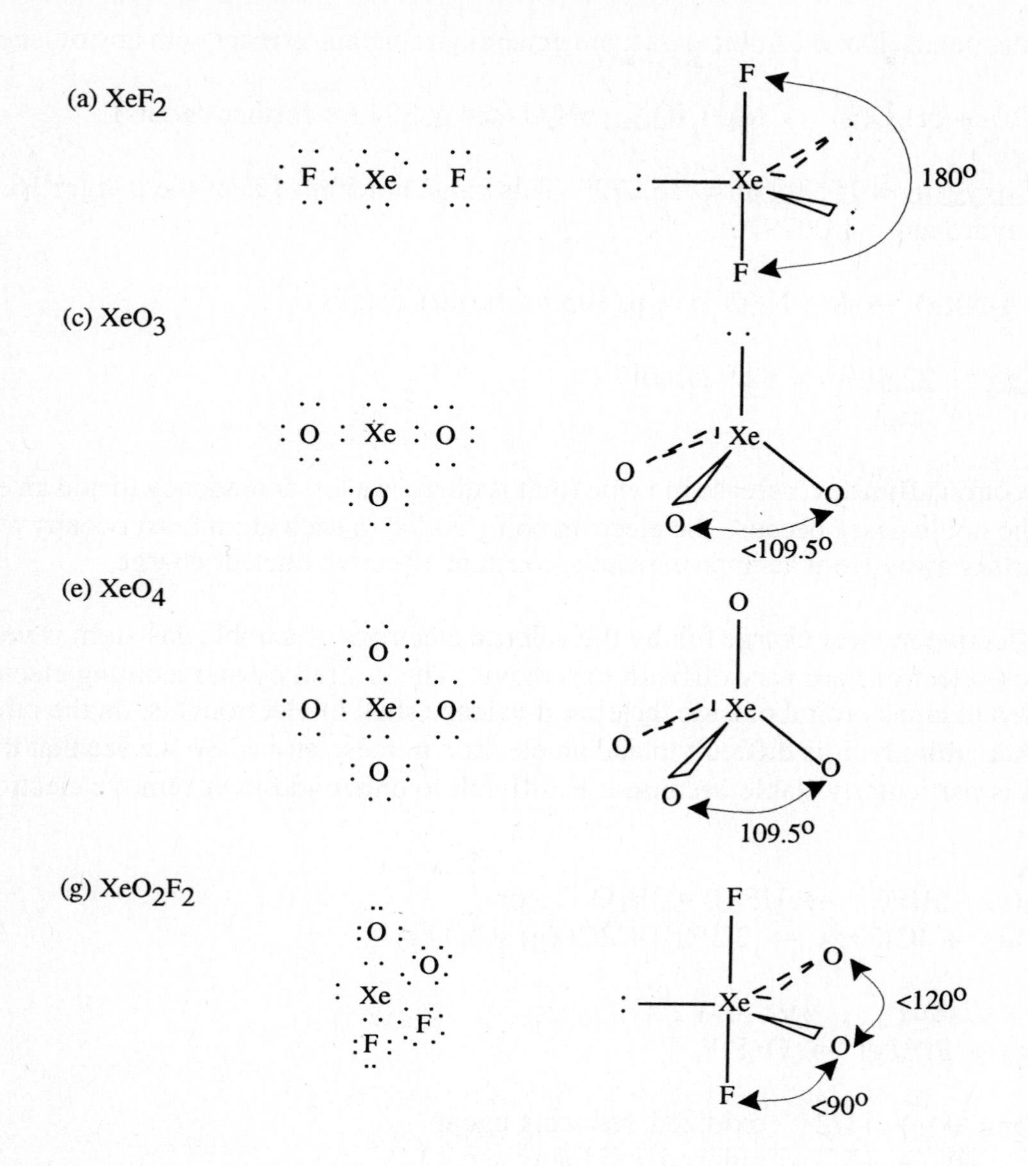

19.29.

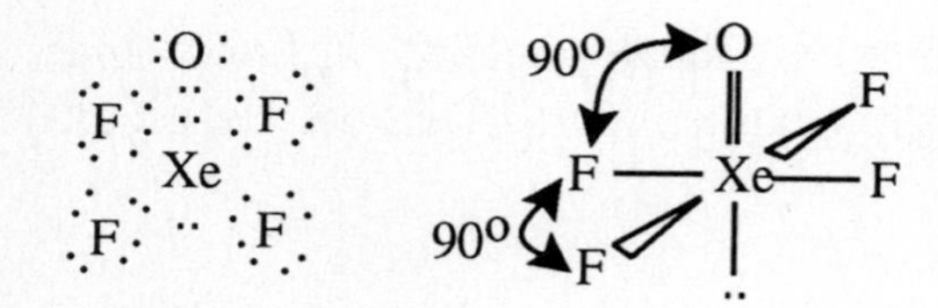

The F-Xe-O angle may be approximately 90° because the effect of the lone pair (which works to compress the F-Xe-O bond) is approximately balanced by the X=O double bond (which works to expand the same bond angle). Also, the xenon atom is very large and therefore there is plenty of room for the lone pair to spread out without unduly influencing the Xe-F bonds.

19.31. XeF_6 $XeOF_4$ XeO_2F_2 XeO_3

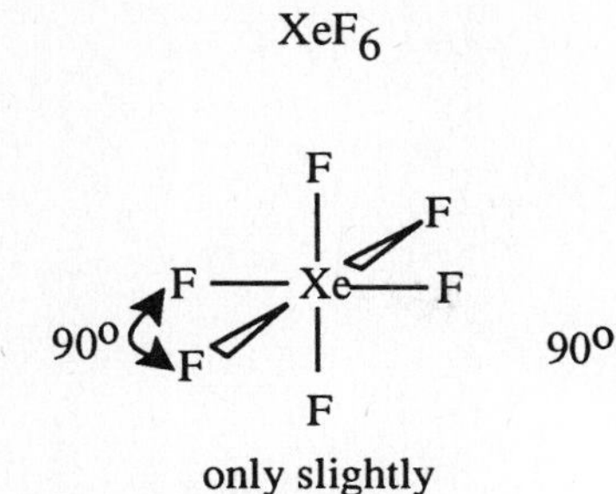

only slightly distorted

90° O F F Xe F 90° F

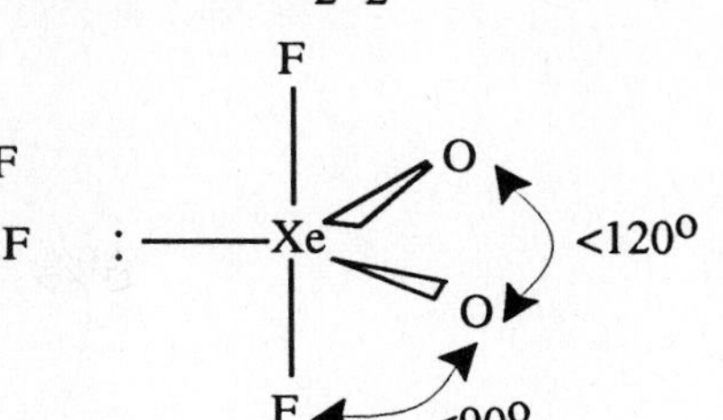

Xe O O O <109.5°

19.33. $XeOF_2$

(There may also be some double bond character to the Xe-O bond.)

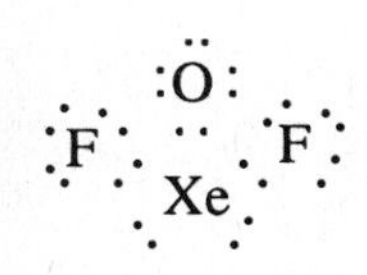

F <90° O Xe F

19.35. $K^{129}_{53}ICl_4 \rightarrow 2\ ^{0}_{-1}e + K^+ + ^{129}_{54}XeCl_4$

Cl Cl Xe Cl 90° Cl

19.37. (a) $XeO_6^{4-}(aq) + 12H^+(aq) + 8e^- \rightarrow Xe(g) + 6H_2O$

(b) $5\ [XeO_6^{4-}(aq) + 12H^+(aq) + 8e^- \rightarrow Xe(g) + 6H_2O]$

$8\ [4H_2O + Mn^{2+}(aq) \rightarrow MnO_4^-(aq) + 8H^+ + 5e^-]$

$5XeO_6^{4-}(aq) + 2H_2O + 8Mn^{2+}(aq) \rightarrow 5Xe(g) + 8MnO_4^-(aq) + 4H^+(aq)$

19.39. $^{40}_{19}K \xrightarrow{\beta^+} ^{40}_{18}Ar + ^{0}_{+1}e$

19.41. Radon gas seeps from the soil around houses into basements through cracks in floors and walls. Once in the basement, radon tends to collect there because it has a density (9.73 g/L) so much greater than air.

19.43. What did you find? Is radon still considered to be a national environmental health problem? Have other geographical areas with high levels of radon been discovered? Is the problem still pretty much confined to private homes? What about public buildings or even caves? Are radon test kits still readily available in department stores?